Angular Momentum

THIRD EDITION

Angular Momentum

THIRD EDITION

D. M. BRINK
Università degli Studi di Trento

and

G. R. SATCHLER
Oak Ridge
National Laboratory

CLARENDON PRESS · OXFORD

Oxford University Press, Walton Street, Oxford OX2 6DP
Oxford New York
Athens Auckland Bangkok Bombay
Calcutta Cape Town Dar es Salaam Delhi
Florence Hong Kong Istanbul Karachi
Kuala Lumpur Madras Madrid Melbourne
Mexico City Nairobi Paris Singapore
Taipei Tokyo Toronto
and associated companies in
Berlin Ibadan

Oxford is a trade mark of Oxford University Press

Published in the United States
by Oxford University Press Inc., New York

First published 1993
Reprinted 1994

A catalogue record for this book is available from the British Library.

Library of Congress Cataloging in Publication Data
Brink, D. M. (David Maurice)
Angular momentum/D. M. Brink and G. R. Satchler
1. Angular momentum (Nuclear physics) 2. Quantum theory
I. Satchler, G. R. (George Raymond) II. Title
QC793.3.A5B75 1993 539.7'25–dc20 93-32190
ISBN 0 19 851759 9 (Pbk)

Printed and bound in Great Britain by
Biddles Ltd, Guildford and King's Lynn

PREFACE TO THE THIRD EDITION

WE regard this book as an introductory text, so we did not feel any need for major modifications for this new edition. However, the broadened use of semiclassical techniques in quantum mechanics, in both atomic and nuclear physics, prompted us to add Appendix VII which introduces the reader to asymptotic, or 'large-j', approximations for many of the quantities discussed here. These relations provide some insights into the transition from quantal to classical descriptions of the angular momentum aspects of various phenomena, as well as often providing ways of simplifying computations that involve angular momenta that are large compared to Planck's constant $\hbar$. Such large-j expressions can be quite accurate, even if the j-values are not very large, and their use can provide classical insights into the workings of complex phenomena. In other cases, large-j relations are valid only if some averaging over quantal oscillations is performed, emphasizing that the transition from quantal to classical descriptions need not converge uniformly.

We also took the opportunity to correct the few misprints remaining in the second edition.

A number of other, more detailed, books pertaining to this subject have appeared since our second edition was published in 1968. A nice treatment by R. N. Zare, *Angular momentum: understanding spatial aspects of chemistry and physics* (Wiley, New York, 1988) provides many examples of applications in atomic and molecular physics. Another text by I. Lindgren and J. Morrison, *Atomic many-body theory* (Springer, Berlin, 1982) makes use of angular momentum theory in detailed applications of the theory of atomic structure. In particular, they use the graphical techniques discussed here in Chapter VII, but they find it physically more transparent to reverse our convention for the direction of the arrow on each leg of the graphs. We might also mention *Angular momentum in quantum physics* and *The Racah–Wigner algebra in quantum theory*, both by L. C. Biedenharn and J. D. Louck (Addison-Wesley, Reading, Massachusetts, 1981), and *Irreducible tensor methods* by B. L. Silver (Academic Press, New York, 1976). Last, but not least, an English translation of the very extensive compendium of formulae and relationships *Quantum theory of angular momentum* by D. A. Varshalovich, A. N. Moskalev, and V. K. Khersonskii has been published (World Scientific, Singapore, 1988). Other useful results and applications are given in refs. [88–95].

May 1993
Italy and Oak Ridge

D.M.B.
G.R.S.

PREFACE TO THE SECOND EDITION

SINCE publication of the first edition, a number of books have appeared that treat various aspects of the quantum theory of angular momentum. One, by Yutsis, Levinson, and Vanagas (Israel Programme for Scientific Translations, Jerusalem, 1962), develops the graphical methods introduced by Levinson. We have added a new chapter which gives a brief introduction to graphical techniques. We might also mention *Quantum Theory of angular momentum*, edited by C. L. Biedenharn and H. Van Dam (Academic Press, New York, 1965) which reprints a number of important papers in the field and includes an extensive bibliography of basic theory and applications up to 1965. In particular, detailed applications to the shell theory of nuclear structure have been given by A. de Shalit and I. Talmi, *Nuclear shell theory* (Academic Press, New York, 1963), while B. R. Judd has described *Operator techniques in atomic spectroscopy* (McGraw-Hill, New York, 1963). Additional tables of numerical values for the various coefficients have now appeared; we might mention the *Tables of Racah coefficients* by A. F. Nikiforov, V. B. Uvarov, and Yu. L. Levitan (Macmillan, New York, 1965), and *A table of Clebsch–Gordan coefficients* by B. E. Chi (Report prepared by Rensselaer Polytechnic Institute, Troy, New York, 1962).

Errors and misprints found in the first edition have been corrected. There are only two changes of notation: The Fano X-coefficient has been replaced by the Wigner 9-j symbol and the definition of certain quantities in section 6.1.2 has been changed to correct and clarify equation (6.21). The Wigner 6-j symbol has come to be used more frequently and formulae involving this function have been collected in Appendixes II and VI. A few other useful formulae have been added to the appendixes.

We would like to thank many people who have brought various errors and misprints to our attention. We are especially indebted to Dr. K. T. R. Davies, Dr. H. J. Rose, Dr. O. Häusser, and Dr. J. Lopes for checking the text and many of the formulae.

D.M.B.

March 1967 G.R.S.

PREFACE TO THE FIRST EDITION

DURING recent years important technical advances have been made in the quantum theory of angular momentum and its application to physical problems, both in nuclear and atomic physics. Our intention is to present these new techniques and to explain their physical significance without undue reference to their highly formal group-theoretic origins. Since this book was started other texts have been published by Rose [54], Edmonds [22] and Messiah [46]. Some overlap with their work is inevitable, but we feel our approach is sufficiently different to make a useful contribution. We have, in particular, attempted to emphasize the physical applications and to provide a source of formulae for workers in this field. Much of the underlying formal theory was developed by Wigner as early as 1937 [77], and is discussed in the book by Fano and Racah [31] and the well-known text of Wigner recently translated into English, [78].

References made to the literature on the physical applications of this theory are necessarily selected somewhat arbitrarily. We can hope to do no more than provide a starting point for wider reading. Where possible reference has been made to review articles rather than individual papers. We beg the indulgence of any authors whose work does not seem to be given proper recognition.

Since there is a bewildering variety of notations and phase conventions it may be a help to the reader to have an indication of some of those adopted in this book. Others can be found in the relevant sections.

We use **J** and **L** to denote angular momentum vectors. Generally **L** refers to orbital angular momentum, but there is no fixed rule. In order to simplify formulae angular momentum is measured for the most part in natural units corresponding to $\hbar = 1$. For spherical harmonics Y_{kq} we adopt the usual phase convention of Condon and Shortley [17] and make frequent use of the 'renormalized' spherical harmonics

$C_{kq} = [(2k + 1)/4\pi]^{-\frac{1}{2}} Y_{kq}$ which help avoid unnecessary factors of 4π in many formulae. There are many notations for Clebsch–Gordan coefficients (cf. Appendix I) but fortunately most definitions agree. We often use the more symmetric Wigner 3-j symbol. There are two main definitions for rotation matrices and ours is explained in sections 1.4 and 2.4 and in Appendix V.

D.M.B.

November 1961 G.R.S.

CONTENTS

CHAPTER I

SYMMETRY IN PHYSICAL LAWS

1.1. Introduction

A THEORETICAL investigation of a physical problem may come upon difficulties of two kinds. The exact physical laws governing the behaviour of a system may not be known, making it impossible to arrive at a complete theoretical description of the system. Problems of fundamental particle structure and reactions present this first difficulty in an acute form. More often the situation is analogous to that encountered in problems of atomic and molecular structure, where the interactions (Coulomb forces) are well-known, but where the structure in a particular problem may be so complicated that no exact solution can be found.

Fortunately the basic interactions in most physical problems have symmetry properties which affect the structure of a composite system in a way independent of the details of the interactions. Symmetries of physical laws may lead to conservation laws, and the laws of conservation of energy, momentum, angular momentum and isotopic spin arise in this way. Again, in the theory of molecular structure the symmetry of the configuration of nuclei in a molecule produces a symmetry in the electronic structure. An understanding of the effects of symmetry often enables one to distinguish between properties of a physical system which are consequences of conservation laws, and properties depending upon details of structure and interaction. For example, the angular distribution between two radiations emitted successively from a nucleus depends partly on symmetry properties, i.e. on the angular momenta of the states involved and partly on the detailed structure of the states. An understanding of the dependence on symmetry of the interactions enables one to obtain information about the detailed structure from experiments. A more familiar classical example is given by the motion of a particle in a central field. The symmetry of the interaction leads to a plane orbit and conservation of angular momentum while the exact shape of the orbit depends upon the detailed form of the central interaction.

1.2. The Symmetry of Physical Laws

We develop the ideas suggested in the introduction by defining a symmetry transformation of a dynamical system as a transformation which applied to any dynamically possible state of motion carries it into another possible state of motion. As an example consider the solution to a classical problem involving two particles expressed by giving the positions of the particles as functions of the time. If the two particles are identical in their intrinsic properties and interactions then interchanging the trajectories of the particles yields a second solution to the problem. In the sense of the above definition the operation of interchanging two particles is a symmetry transformation.

In quantum mechanics a symmetry transformation has a very simple representation. A state of motion of a quantum system is determined by giving the wave function $\psi(t)$ as a function of time. This wave function represents a possible state of motion if the Schrödinger equation is satisfied;

$$i\hbar\frac{d\psi}{dt} = H\psi.$$

A symmetry transformation is represented by a linear operator U acting on the wave function with the property that $U\psi(t)$ satisfies the Schrödinger equation whenever it is satisfied by $\psi(t)$, i.e. if $\psi(t)$ represents a possible state of motion then so does $U\psi(t)$. Symmetry transformations preserve the orthogonality of wave functions and, in general, can be represented by unitary operators‡, so that $U^{-1} = U^{+}$. If the symmetry

‡ A unitary operator satisfies the condition $U^{+}U = UU^{+} = 1$, where U^{+} is the adjoint of the operator, i.e. the adjoint operator U^{+} is also the inverse to U. A unitary transformation preserves the orthogonality and normalization of wave-functions. Some symmetry transformations are represented by anti-unitary operators, cf. section 1.7.

transformation U is time independent then

$$i\hbar \frac{dU\psi}{dt} = HU\psi$$

$$\text{hence } i\hbar \frac{d\psi}{dt} = U^{-1}HU\psi$$

$$\text{and } UH = HU.$$

Thus the operator representing a time independent symmetry transformation commutes with the Hamiltonian of the system.

1.3. The Symmetry Group of a Dynamical System

An abstract group is characterised by the following properties. It is a set of elements a, b . . . with a multiplication law defined satisfying the conditions that

(1) it is associative $a(bc) = (ab)c$,
(2) there is a unit element 1 such that $1a = a1 = a$,
(3) every element a has an inverse a^{-1} with the property that $aa^{-1} = a^{-1}a = 1$, and a^{-1} is itself an element of the group.

The group multiplication law is often non-commutative so that $ab \neq ba$. A sub-group of a given group is a sub-set of the group elements which itself satisfies the conditions of a group, in particular that the product of any two elements of the sub-group must lie within the sub-group, and the inverse of an element of the sub-group must lie in the sub-group.

The set of all non-singular square matrices of order n with the matrix multiplication law provides a particular example of a group. Conditions (1) to (3) are satisfied since matrix multiplication is associative, there is a unit matrix of order n, and every non-singular square matrix has an inverse. Also the product of two non-singular matrices is non-singular. This group is called the full linear group of order n. The unitary group of order n is the set of all $n \times n$ matrices A with the property that $A^{-1} = A^{+}$ where A^{+} is the adjoint matrix to A. It forms a sub-group of the full linear group of order n.

Returning to symmetry transformations it is clear that the successive application of two symmetry transformations produces a third symmetry transformation, and it can be shown that the set of all symmetry transformations with this law of combination forms a group, the symmetry group of the system. For a quantum system the symmetry transformations are represented by unitary operators, and the successive application of two symmetry transformations S and T is represented by the operator product TS. From section 1.2 we see the set of symmetry transformations of a quantum system is contained in the set of all unitary operators which commute with the Hamiltonian operator of the system.

1.4. Geometrical Symmetries

An important sub-group of symmetry transformations of many dynamical systems has a geometrical origin. The space of classical physics is described by a Euclidean geometry, implying that all points of space and all directions are equivalent and only statements relating to relative position and relative orientation have a meaning independent of the coordinate system. If the Euclidean character of space is reflected in physical laws no point in space nor any directions should be distinguishable absolutely by the performance of experiments. The operations of translation and rotation applied to a physical system should therefore belong to the symmetry group of the system. Thus the geometrical character of space determines symmetries of physical laws, or, perhaps more correctly, the physical symmetries determine the geometry of space. Lorentz transformations and space reflection come also into the category of symmetries with a geometrical origin.

In practical problems often a distinction is made between a physical system and its surroundings and the effect of the surroundings is approximated by boundary conditions or by a set of known fields or forces applied from the outside. This approximate treatment may destroy the geometrical symmetries partially or completely. For example, the equations of motion of an atom in an external magnetic field are not invariant for arbitrary rotations unless the external field is also rotated.

The geometrical symmetry operations of translation and rotation can be looked at in two ways, called active and passive. In the active sense introduced above they transform one state of a system into another. A rotation actually rotates the system from one position to another. The passive approach

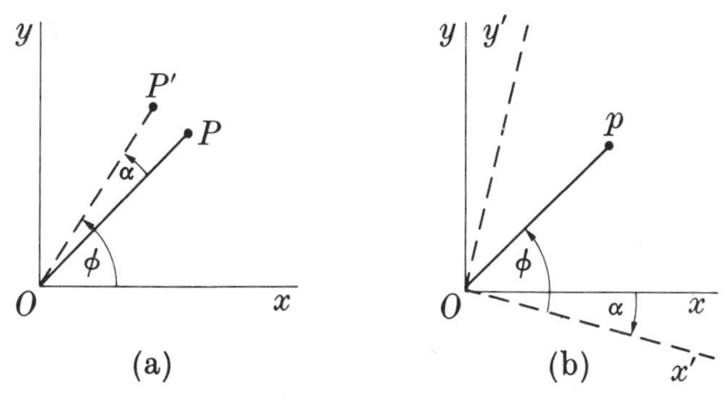

FIG. 1. (a) Represents a rotation of the system. A vector OP is carried into OP' by the rotation.
(b) Represents the equivalent rotation of axes. The vector OP remains fixed, but its coordinates are changed. The position of OP relative to the new axes in (b) is the same as that of OP' relative to the axes in (a).

interprets the symmetry operations as coordinate transformations, and the symmetry of the dynamical laws is expressed by stating that the equations of motion are invariant under translation or rotation of axes. Any translation or rotation of a system may, however, be induced by a coordinate transformation and the two views are equivalent. This fact can be seen most easily by considering a simple example, the rotation of a one-particle quantum system through an angle α about the z-axis. (See Fig. 1.) Rotating the system transforms its wave function to a new one so that the value of the new wave function at the point $(\phi+\alpha)$ is the same as that of the old one at the point ϕ where ϕ is the azimuthal angle. Let the original wave function be $\psi(r,\theta,\phi)$ and the rotated wave function

$\psi' = D_\alpha \psi$ then this argument shows that

$$\psi'(r,\theta,\phi+\alpha) = \psi(r,\theta,\phi),$$

or

$$\psi'(r,\theta,\phi) = \psi(r,\theta,\phi-\alpha).$$

Thus the rotated wave function can be obtained from the original wave function by making the coordinate transformation $(r,\theta,\phi) \rightarrow (r,\theta,\phi-\alpha)$, i.e. by a rotation of the coordinate axes through an angle $-\alpha$ about the z-axis. Both these views of the rotation are important and will be used with the convention that a positive (counter clockwise) rotation refers to a positive rotation of the system, hence a negative rotation of axes.‡

1.5. Conservation Laws

There is a relation between the geometrical symmetries of a physical system and the momentum and angular momentum conservation laws, having its origin in the fact that the linear and angular momenta are respectively the generators of translations and rotations of the system. As an example consider the canonical transformation generated by L_z, the z-component of the angular momentum of a classical particle.

If F is any function of the coordinates q_i and the momenta p_i of a dynamical system and α is small then the transformation§

$$\begin{aligned} p_i' &= p_i+\alpha\{p_i,F\}, \\ q_i' &= q_i+\alpha\{q_i,F\}, \end{aligned} \tag{1.1}$$

is an infinitesimal canonical transformation in the sense that Hamilton's equations of motion retain the same form when written in terms of p_i' and q_i'. F is called the generating function of the transformation and it can be shown that if $G(p_i, q_i)$

‡ This convention agrees with Rose [54], Messiah [46]. Other authors use the opposite convention, a positive rotation referring to a positive rotation of the axes; Wigner [78], Edmonds [22]. Cf. section 2.4 and Appendix V.

§ The Poisson bracket $\{G, F\}$ of two functions of p_i and q_i is defined as

$$\{G, F\} = \sum_k \left(\frac{\partial G}{\partial q_k}\frac{\partial F}{\partial p_k} - \frac{\partial G}{\partial p_k}\frac{\partial F}{\partial q_k}\right).$$

is any function of the coordinates and momenta then the same function of the transformed coordinates is

$$G(p_i',q_i') = G(p_i,q_i)+\alpha\{G,F\}. \tag{1.2}$$

The orbital angular momentum of a particle is the vector product $\mathbf{L} = \mathbf{r} \wedge \mathbf{p}$ and in particular $L_z = xp_y - yp_x$. If we replace p_i and q_i in equation (1.1) by the rectangular coordinates and momenta of the particle and F by L_z then simple calculation shows that the infinitesimal transformation induced is

$$\begin{aligned} x' &= x-\alpha y, & p_x' &= p_x-\alpha p_y, \\ y' &= y+\alpha x, & p_y' &= p_y+\alpha p_x, \\ z' &= z, & p_z' &= p_z. \end{aligned} \tag{1.3}$$

It corresponds to a rotation of the particle position and momentum through an infinitesimal angle α about the z-axis. The angular momentum component L_z generates infinitesimal rotations about the z-axis.

In classical dynamics the rate of change of any function F of coordinates and momenta with time is given by the equation

$$\frac{dF}{dt} = \{F,H\} \tag{1.4}$$

with $H(p,q)$ the Hamiltonian of the system. Invariance of the equations of motion under rotations implies that the Hamiltonian should be unchanged by a rotation

$$H(\mathbf{r}',\mathbf{p}') = H(\mathbf{r},\mathbf{p}). \tag{1.5}$$

Transformation theory (equation (1.2)) requires, however, that

$$H(\mathbf{r}',\mathbf{p}') = H(\mathbf{r},\mathbf{p})+\alpha\{H,L_z\}$$

implying that $$\{L_z,H\} = 0,$$

and from equation (1.4) $$\frac{dL_z}{dt} = 0.$$

Thus the motion is such that L_z is constant in time, so we see that invariance under rotations about an axis (equation (1.5)) implies conservation of angular momentum about that axis.

Similar considerations establish the connection between translation invariance of the equations of motion and conservation of linear momentum. The laws of conservation of momentum and angular momentum are closely connected with the geometrical symmetries of translation and rotation, hence in a sense they are geometrical in origin.

The above discussion has been given in terms of a simple classical system, but the results are general and hold also in quantum mechanics. The derivation follows an identical pattern in view of the close analogy between classical dynamics expressed in terms of Poisson brackets and quantum mechanics in the Heisenberg representation, reflected in the correspondence between the Poisson bracket and the quantum commutator

$$FG-GF = [F,G] = i\hbar\{F,G\}. \tag{1.6}$$

F and G are functions of the coordinates and momenta in the classical case and of the corresponding operators in quantum mechanics.

A transformation D applied to a quantum system can be interpreted according either to the Schrödinger or the Heisenberg representation. If A is an observable then results of observations correspond to matrix elements of A, $\langle 1|A|2\rangle$ between states of the system. The corresponding matrix element in the transformed system is $\langle 1|D^+AD|2\rangle$ and the transformation can be interpreted either as a transformation of the state vectors $|1\rangle \rightarrow D|1\rangle$ and $|2\rangle \rightarrow D|2\rangle$ or as an operator transformation $A \rightarrow A' = D^+AD$ leaving the state vectors unchanged. The second interpretation is more appropriate to the Heisenberg representation and we shall use it to derive an expression for the operator D, which rotates a quantum system, from the classical results of equations (1.2) and (1.3) and the correspondence principle expressed in the Poisson bracket relation (1.6). When a system is rotated through an infinitesimal angle α about the z-axis an operator A transforms to A' according to equation (1.2)

$$A' = A+\alpha\{A,L_z\}$$

to first order in α. The correspondence relation (1.6) implies

$$A' = A - \frac{i\alpha}{\hbar}(AL_z - L_z A),$$

$$= (1 + \frac{i\alpha}{\hbar}L_z)A(1 - \frac{i\alpha}{\hbar}L_z).$$

Thus the operator D_α for an infinitesimal rotation about the z-axis is

$$D_\alpha = (1 - \frac{i\alpha}{\hbar}L_z).$$

By integrating the operator for infinitesimal rotations it can be shown that the operator for rotation through a finite angle α about the z-axis has the explicit form

$$D_\alpha = \exp(-i\alpha L_z/\hbar). \tag{1.7}$$

In the following, to simplify formulae, we suppose that angular momentum is measured in natural units so that $\hbar = 1$.

Some quantum systems have non-classical internal degrees of freedom ('spin' degrees of freedom) in addition to the classical ones. The orbital angular momentum operator $\mathbf{L} = \sum_i \mathbf{r}_i \wedge \mathbf{p}_i$ generates rotations of the classical variables, while the spin angular momentum operator $\mathbf{S} = \sum_i \mathbf{S}_i$ rotates the internal degrees of freedom. The generator for rotations of the system as a whole is the total angular momentum $\mathbf{J} = \mathbf{L}+\mathbf{S}$ and $\mathbf{J}$ rather than $\mathbf{L}$ or $\mathbf{S}$ separately is conserved as a result of invariance under rotations. If $\mathbf{L}$ and $\mathbf{S}$ happen to be conserved separately, as is approximately the case in some atoms, it is a specifically physical rather than geometrical property.

1.6. Commutation Rules for J

Let $D_\alpha = 1 - i\alpha J_z$ be the operator which rotates a system through an infinitesimal angle α about the z-axis. This rotation applied to a vector operator $\mathbf{A}$ rotates it through an angle α about the z-axis. The components of the rotated vector $\mathbf{A}'$ are

$$A'_x = A_x - \alpha A_y,\ A'_y = A_y + \alpha A_y,\ A'_z = A_z \tag{1.8}$$

From the discussion of section 1.5, however, A_x transforms to

$$\begin{aligned} A'_x &= D^+_\alpha A_x D_\alpha, \\ &= (1+i\alpha J_z)A_x(1-i\alpha J_z), \\ &= A_x+i\alpha(J_z A_x - A_x J_z). \end{aligned} \tag{1.9}$$

Comparison of equations (1.8) and (1.9) yields

$$J_z A_x - A_x J_z = [J_z, A_x] = iA_y,$$

and similarly

$$[J_z, A_y] = -iA_x,$$

and

$$[J_z, A_z] = 0.$$

Similar commutation relations are obtained with J_x and J_y by rotation about the x- and y-axes respectively. If **A** is the angular momentum vector **J** itself we get

$$\begin{aligned} J_x J_y - J_y J_x &= iJ_z, \\ J_y J_z - J_z J_y &= iJ_x, \\ J_z J_x - J_x J_z &= iJ_y, \end{aligned} \tag{1.10}$$

or

$$\mathbf{J} \wedge \mathbf{J} = i\mathbf{J},$$

for the commutation relations of the components of **J**. These commutation relations also follow directly for the orbital angular momentum $\mathbf{L} = \mathbf{r} \wedge \mathbf{p}$ from the commutation relations of **r** with **p**.

1.7. Parity

Reflection through the origin $x \to -x$, $y \to -y$, $z \to -z$ is a third symmetry operation of a geometrical and therefore 'intuitive' nature. It differs from the operations of translation and rotation in that it is discontinuous. Classically this implies that invariance under reflections leads to no conservation law, in the way that invariance under rotations leads to conservation of angular momentum. This is not the case in quantum mechanics. If P is the operator reflecting a system through the origin and P is a symmetry operation then by section 1.5 in the Heisenberg representation

$$i\hbar\frac{\partial P}{\partial t} = [P, H] = PH - HP = 0,$$

and the operator P is constant in time, or 'parity' is conserved. It has been found, however, that the interaction Hamiltonian responsible for β-decay is not invariant under coordinate inversion and so does not commute with P. This may be true also of some other interactions so that parity would not be conserved for those interactions.

1.8. Time Reversal

Another symmetry operation often occurring in conjunction with rotational symmetry is the time reversal transformation [Wigner 78].

$$t \to -t.$$

Time reversal transforms other dynamical variables as follows:

$$\mathbf{r} \to \mathbf{r};\ \mathbf{p} \to -\mathbf{p};\ \mathbf{J} \to -\mathbf{J}.$$

A stationary state of a quantum system has a simple time dependence proportional to $\exp(-iEt/\hbar)$, where E is the energy of the state. The time reversal transformation changes the sign of t in this exponential factor but has no effect on the main part of the wave function. For this reason invariance under time reversal gives no conservation law and no additional quantum numbers. On the other hand time reversal does say something about non-stationary processes leading, for example, to the law of detailed balance for nuclear reactions, [Blatt and Weisskopf 9].

As a further introduction to the ideas of time reversal consider a particle moving in one dimension. If the Hamiltonian of the particle is a function of the coordinate x and the momentum p then invariance under time reversal is expressed by the equation

$$H(x, -p) = H(x, p).$$

Alternatively H is a real operator when expressed in terms of x and $p = -i\hbar\frac{\partial}{\partial x}$. If $\phi(x, t)$ is a solution of the Schrödinger equation for the system then $\phi(x, -t)$ is a solution of the time reversed Schrödinger equation, but since H is real the complex conjugate $\phi^*(x,t)$ is also a solution. If the phase of ϕ is chosen so that ϕ is real when $t = 0$ these two solutions have the same initial value at $t = 0$, and it follows that the solutions are identical for all times

$$\phi(x, -t) = \phi^*(x,t). \tag{1.11}$$

If ϕ is a stationary solution we can write

$$\phi(xt) = \phi_0(x)e^{-iEt/\hbar}.$$

Equation (1.11) shows that the phases can be chosen so that $\phi_0(x)$ is real, hence stationary states can be represented by real wave functions in the coordinate representation. It can be shown further that a complete set of wave functions (with definite phases) can always be found so that all matrix elements of operators invariant under time reversal are real. If we introduce a time reversal operator $\boldsymbol{\theta}$ for the above system by

$$\boldsymbol{\theta}\phi(x,t) = \phi^*(x,t)$$

then
$$\boldsymbol{\theta}(\phi_1+\phi_2) = \phi_1^*+\phi_2^* = \boldsymbol{\theta}\phi_1+\boldsymbol{\theta}\phi_2 \tag{1.12}$$

and
$$\boldsymbol{\theta}(a\phi) = a^*\phi^* = a^*\boldsymbol{\theta}\phi,$$

if a is a complex number. An operator which satisfies equations (1.12) is called anti-linear. A real wave function is invariant under this transformation.

The results found in this special example can be generalized to apply to an arbitrary quantum system invariant under time reversal. A time reversal operator can be defined which is always anti-linear, but it cannot always be represented by simple complex conjugation. Suppose one can find a complete set of wave functions $\{|m\rangle\}$ invariant under the time reversal operation (thus with definite phases). If $|\alpha\rangle$ is an arbitrary vector which is invariant under time reversal then

$$\boldsymbol{\theta}|\alpha\rangle = |\alpha\rangle;$$

but $|\alpha\rangle$ and $\boldsymbol{\theta}|\alpha\rangle$ can be expanded in the complete (invariant) set $\{|m\rangle\}$.

$$|\alpha\rangle = \sum_m |m\rangle\langle m|\alpha\rangle \tag{1.13}$$

and
$$\begin{aligned}\boldsymbol{\theta}|\alpha\rangle &= \sum_m \boldsymbol{\theta}(|m\rangle\langle m|\alpha\rangle)\\ &= \sum_m (\boldsymbol{\theta}|m\rangle)\langle m|\alpha\rangle^* \text{ (antilinear property of } T)\\ &= \sum_m |m\rangle\langle m|\alpha\rangle^*. \ (|m\rangle \text{ is invariant)}\end{aligned} \tag{1.14}$$

Comparing equations (1.13) and (1.14) we see that if $|\alpha\rangle$ is invariant the expansion coefficients $\langle m|\alpha\rangle$ are real. A similar calculation shows that the matrix elements of any operator invariant under time reversal, between states of the set $|m\rangle$ are real. Thus only one real parameter instead of two is required to specify the value of a matrix element. Special problems arise in systems with angular momentum since the angular momentum operator is not invariant; but changes sign on time reversal. A further discussion of this point is given in section 4.9.

CHAPTER II

REPRESENTATIONS OF THE ROTATION GROUP

2.1. Group Representations in Quantum Mechanics

AN important part of the theory of groups is that concerned with the representation of their elements by matrices. If G is an abstract group then a representation of G with dimension n is a correspondence between the elements of G and a subset of the matrices of order n such that to each element a of the group G there is an $n \times n$ matrix $R(a)$ with the property that if a and b are group elements then

$$R(a)R(b) = R(ab),$$

i.e. group multiplication corresponds to matrix multiplication.

In a quantum mechanical formalism the elements of the symmetry group G of a system with Hamiltonian H are represented by unitary operators in the Hilbert space of state vectors or wave functions, all of which commute with H (section 1.2). If E is an n-fold (n-finite) degenerate eigenvalue of H, there exists an n-dimensional manifold $\mathcal{M}$ in the Hilbert space such that all the state vectors in $\mathcal{M}$ are eigenvectors of H with eigenvalue E. Let ϕ be any vector in $\mathcal{M}$ and S any symmetry transformation then the operator S commutes with H and

$$HS\phi = SH\phi = S(E\phi) = ES\phi.$$

Thus $S\phi$ is also an eigenvector of H with eigenvalue E, and $S\phi$ is also in the manifold $\mathcal{M}$: S transforms $\mathcal{M}$ into itself. Choosing $|1\rangle \ldots |n\rangle$ as a set of orthonormal state vectors spanning $\mathcal{M}$, then the transformation of $\mathcal{M}$ by the operator S is represented by the matrix $S_{ij} = \langle i|S|j\rangle$. For if $|a\rangle$ is any state in $\mathcal{M}$ then $|a\rangle$ and $S|a\rangle$ can be expanded in the orthonormal set as $|a\rangle = \sum_i a_i|i\rangle$ and $S|a\rangle = \sum_i b_i|i\rangle$; but

$$S|a\rangle = \sum_i a_i S|i\rangle = \sum_{ij} a_i|j\rangle\langle j|S|i\rangle.$$

Therefore

$$b_j = \sum_i \langle j|S|i\rangle a_i.$$

These matrices form a representation of the symmetry group G of the quantum system, for if S and T are in G then

$$\langle i|ST|j\rangle = \sum_k \langle i|S|k\rangle\langle k|T|j\rangle,$$

and the matrices have the same multiplication law as the group elements they represent. Hence follows the important result that to every n fold degenerate eigenvalue of the Hamiltonian of a system there corresponds an n-dimensional representation of the symmetry group of the system. The representations of the symmetry group can be used to classify degenerate states of the Hamiltonian.

There are an infinite number of ways of choosing the basis of the manifold $\mathcal{M}$ in the above discussion, and for each choice of basis the symmetry transformations are represented by different matrices. These representations are simply related and one can pass from one to another by a change of basis or by a unitary transformation. Representations differing only by a unitary transformation are considered the same.

2.1.1. Reduction of a Representation

Consider a representation Δ of a group G on a manifold $\mathscr{M}$. Suppose there exist sub-manifolds $\mathscr{M}_1$ and $\mathscr{M}_2$ of $\mathscr{M}$ such that $\mathscr{M}_1+\mathscr{M}_2=\mathscr{M}$ and any matrix of the representation Δ transforms a state in $\mathscr{M}_1$ to a state in $\mathscr{M}_1$ and a state in $\mathscr{M}_2$ to a state in $\mathscr{M}_2$. If such a decomposition is possible then the representation Δ is said to be reducible. Otherwise the manifold $\mathscr{M}$ is irreducible under the operations of the group and the representation is irreducible. If a basis is chosen for the manifold $\mathscr{M}$ so that the vectors $|1\rangle\ldots|r\rangle$ span $\mathscr{M}_1$ and the vectors $|r+1\rangle\ldots|n\rangle$ span $\mathscr{M}_2$ then any matrix T_{ij} of the representation Δ takes the partially diagonalized form

$$T_{ij}=\begin{bmatrix} T_{11} & \cdots & T_{1r} & & & \\ \vdots & & \vdots & & 0 & \\ T_{r1} & \cdots & T_{rr} & & & \\ & & & T_{r+1,r+1} & \cdots & T_{r+1,n} \\ & 0 & & \vdots & & \vdots \\ & & & T_{n,r+1} & \cdots & T_{nn} \end{bmatrix}$$

and the sub-matrices of T_{ij} correspond to representations of the group G of dimension r and $n-r$ respectively. The representation Δ has been reduced to a sum of two representations, Δ_1 of dimension r and Δ_2 of dimension $n-r$. This reduction is written symbolically as

$$\Delta=\Delta_1+\Delta_2$$

2.2. The Irreducible Representations of the Rotation Group

In the following we confine our attention to systems with rotational symmetry and consider only the rotational subgroup of the symmetry group. Consider a finite manifold $\mathscr{M}$

associated with an irreducible representation Δ of the rotation group. Any rotation can be produced by a succession of infinitesimal rotations, thus a necessary and sufficient condition for the irreducibility of Δ is that the manifold $\mathscr{M}$ should be irreducible with respect to the generators J_x, J_y, J_z of infinitesimal rotations.

To find the irreducible representations we introduce the operators $J_{\pm}$ defined by

$$J_{\pm} = J_x \pm iJ_y. \tag{2.1}$$

The operators J_x, J_y, J_z obey the angular momentum commutation laws (1.10), and it follows that

$$J_{\pm} J_z - J_z J_{\pm} = \mp J_{\pm}. \tag{2.2}$$

Let $|j\rangle$ be the eigenvector of J_z with the largest eigenvalue j. Then equation 2.2 gives

$$J_z J_- |j\rangle = J_- J_z |j\rangle - J_- |j\rangle = (j-1) J_- |j\rangle$$

and $J_-|j\rangle$ is an eigenvector of J_z with eigenvalue $j-1$. Similarly $(J_-)^2|j\rangle$ is an eigenvector of J_z with eigenvalue $j-2$ and so on. Let $|j\rangle$, $|j-1\rangle$, ... $|j-r\rangle$ with $|j-r\rangle = (J_-)^r|j\rangle$ be a sequence of eigenvectors of J_z generated by successive application of J_-.

Again $J_z J_+|j\rangle = (j+1)J_+|j\rangle$; but since j is already the largest eigenvalue of J_z in the manifold $\mathscr{M}$, $J_+|j\rangle$ must vanish. The square of the total angular momentum has the expression

$$\begin{aligned}\mathbf{J}^2 &= J_x^2 + J_y^2 + J_z^2 \\ &= J_+ J_- + J_z^2 - J_z \\ &= J_- J_+ + J_z^2 + J_z.\end{aligned} \tag{2.3}$$

It follows that

$$\mathbf{J}^2|j\rangle = (J_- J_+ + J_z^2 + J_z)|j\rangle = j(j+1)|j\rangle,$$

so that $|j\rangle$ is an eigenvector of $\mathbf{J}^2$ with eigenvalue $j(j+1)$. $\mathbf{J}^2$ commutes with J_-, hence $|j-r\rangle$ is also an eigenvector of $\mathbf{J}^2$ with the same eigenvalue $j(j+1)$.

Because $\mathscr{M}$ is a finite manifold the sequence $|j\rangle$ ... $|j-r\rangle$ of orthogonal eigenvectors of J_z generated from $|j\rangle$ must terminate, say, with $r = n$, i.e. $J_-|j-n\rangle = 0$. Now from equation (2.3)

$$\begin{aligned}\mathbf{J}^2|j-n\rangle &= (J_+J_-+J_z^2-J_z)|j-n\rangle,\\ &= (J_z^2-J_z)|j-n\rangle,\\ &= \{(j-n)^2-(j-n)\}|j-n\rangle,\end{aligned}$$

but it was shown that $\mathbf{J}^2|j-n\rangle = j(j+1)|j-n\rangle$,

hence $$j(j+1) = (j-n)^2-(j-n),$$

or $$j = \frac{n}{2}.$$

The number n is a positive integer, hence j must be a positive integer or a positive integer plus one-half. The operators J_+, J_-, J_z transform the vectors $|j\rangle \dots |j-n\rangle$ amongst themselves, and since the manifold $\mathcal{M}$ is irreducible these vectors must span $\mathcal{M}$, so the representation is of dimension $n+1 = 2j+1$. The basis vectors $|j\rangle \dots |j-n\rangle$ are eigenvectors of J_z with eigenvalues ranging in integer steps from $+j$ to $-j$. After normalization we denote these basis vectors by $|j, m\rangle$ where $J_z|j, m\rangle = m|j, m\rangle$ and $-j \leqslant m \leqslant j$. They are all eigenvectors of $\mathbf{J}^2$ with eigenvalues $j(j+1)$ and the non-vanishing matrix elements of $\mathbf{J}$ are given by

$$\begin{aligned}\langle jm|J_z|jm\rangle &= m,\\ \langle jm\pm 1|J_\pm|jm\rangle &= \{(j\pm m+1)(j\mp m)\}^{\frac{1}{2}}.\end{aligned} \tag{2.4}$$

The phases of the off diagonal matrix elements are arbitrary, but normally they are chosen as above [Condon and Shortley 17], thus fixing the relative phases of $|jm\rangle$ and $|jm'\rangle$.

The matrix elements of the infinitesimal rotation operators are determined by the dimension $(2j+1)$ of the representation once the z-axis has been chosen; thus the representation of dimension $(2j+1)$ is unique. All other representations of the same dimension can differ only by a unitary transformation. This representation of dimension $(2j+1)$ is usually denoted by $\mathcal{D}_j$ and corresponds to an eigenvalue $j(j+1)$ of $\mathbf{J}^2$ with j integral or half-odd integral. In particular the basis $|jm\rangle$ of the representation $\mathcal{D}_j$ depends upon the particular choice of the

z-axis. The transformations of the basis corresponding to a change of axes will be discussed in section 2.6.

2.3. Integral Representations and Spherical Harmonics

The representations with j integral arise in problems concerning the orbital angular momentum of a single particle. The state vectors are functions of the particle coordinates, and the angular momentum has the explicit form

$$\mathbf{L} = \mathbf{r} \wedge \mathbf{p}, \quad \text{with} \quad p_x = -i\frac{\partial}{\partial x}, \text{ etc.,}$$

or in polar coordinates,

$$\begin{aligned} L_{\pm} &= L_x \pm iL_y, \\ &= \pm e^{\pm i\phi}\left(\frac{\partial}{\partial\theta} \pm i \cot\theta \frac{\partial}{\partial\phi}\right), \\ L_z &= -i\frac{\partial}{\partial\phi}. \end{aligned} \tag{2.5}$$

Also
$$\mathbf{L}^2 = -\left[\frac{1}{\sin\theta}\frac{\partial}{\partial\theta}\left(\sin\theta\frac{\partial}{\partial\theta}\right) + \frac{1}{\sin^2\theta}\frac{\partial^2}{\partial\varphi^2}\right]. \tag{2.6}$$

If we take eigenfunctions of $\mathbf{L}^2$ and L_z as basis vectors of the irreducible representations then these eigenfunctions are the spherical or surface harmonics,

$$Y_{lm}(\theta\phi) = \Theta_{lm}(\theta)\Phi_m(\varphi),$$

with
$$\begin{aligned} \Theta_{lm}(\theta) &= (-1)^m\left[\frac{(2l+1)}{2}\frac{(l-m)!}{(l+m)!}\right]^{\frac{1}{2}} P_l^m(\theta), \quad \text{if} \quad m \geqslant 0 \\ &= (-1)^m \Theta_{l|m|}(\theta), \quad \text{if} \quad m < 0 \\ \Phi_m(\varphi) &= (2\pi)^{-\frac{1}{2}} e^{im\phi}. \end{aligned}$$

$P_l^m(\theta)(m \geqslant 0)$ is the associated Legendre polynomial [Jahnke and Emde 40]. This definition of the spherical harmonics involves an arbitrary choice of phase and we follow Condon and Shortley [17]. With this choice

$$(Y_{lm}(\theta\varphi))^* = (-1)^m Y_{l-m}(\theta\varphi). \tag{2.7}$$

An alternative phase differing by i^l from the above is used by some authors [7] to give the spherical harmonics a more convenient transformation under time reversal.
When $m = 0$

$$Y_{l0}(\theta,\varphi) = \left(\frac{2l+1}{4\pi}\right)^{\frac{1}{2}} P_l(\cos\theta), \tag{2.8}$$

where P_l (cos θ) is a Legendre polynomial [Jahnke and Emde 40].

Reflection through the origin replaces (θ, ϕ) by $(\pi-\theta, \pi+\phi)$ hence $\cos\theta \rightarrow -\cos\theta$ and it follows from the property of Legendre polynomials

$$P_l^m(\pi-\theta) = (-)^{l-m}P_l^m(\theta),\; e^{im(\pi+\phi)} = (-)^m e^{im\phi}$$

that the spherical harmonics have a definite parity $(-1)^l$ where l is the order of the spherical harmonic. Spherical harmonics are normalized and orthogonal over the unit sphere

$$\int Y_{l'm'}(\theta,\phi)^* Y_{lm}(\theta,\phi)\, d\Omega = \delta(ll')\,\delta(mm'),$$

where $d\Omega = \sin\theta\, d\theta d\phi$ is the infinitesimal element of solid angle, and $\delta(a\,b)$ is unity if $a = b$ and is zero if $a \neq b$. They form a complete set for expanding bounded functions of θ and ϕ.

In some problems it is more convenient to use modified spherical harmonics with a different normalization

$$C_{lm}(\theta,\phi) = \left(\frac{4\pi}{2l+1}\right)^{\frac{1}{2}} Y_{lm}(\theta,\phi). \tag{2.9}$$

With this normalization

$$C_{l0}(\theta,\phi) = P_l(\cos\theta), \tag{2.10}$$

and

$$C_{lm}(0,\phi) = \delta(m\,0). \tag{2.11}$$

2.4. Explicit Representation of the Rotation Matrices

In the previous sections we have found the possible irreducible representations of the rotation group from the commutation properties of the angular momentum operators J_x, J_y, J_z. It remains to discuss the matrices representing finite rotations and for this purpose it is necessary to introduce a set of parameters to specify arbitrary rotations. These parameters are the Euler angles (α, β, γ). If a set of orthogonal axes (x, y, z) are rotated to new positions x', y', and z', the Euler angles are defined as follows (Fig. 2) by making the rotation in three

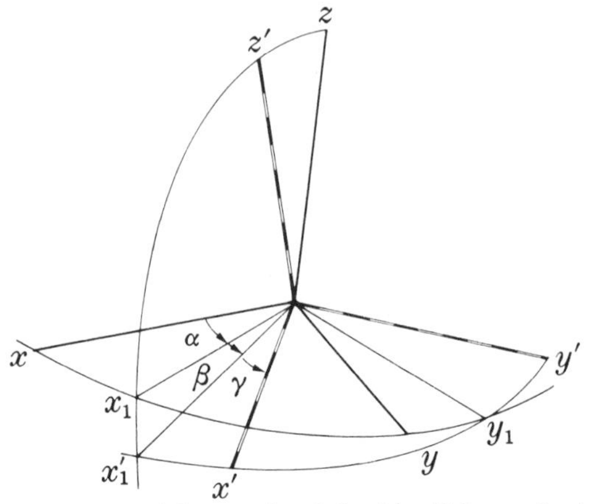

FIG. 2. Illustration of the rotation defined by Euler angles (α,β,γ).

steps. First transform the axes to new positions (x_1, y_1, z) by rotating through an angle α about the z-axis. Next rotate through an angle β about the y_1-axis to positions (x_1', y_1, z') and finally make a rotation through an angle γ about the z'-axis. Positive rotations are defined by the right hand screw sense.

From equation (1.7) the rotation operator corresponding to this rotation has the explicit form

$$D(\alpha, \beta, \gamma) = \exp(-i\gamma J_{z'})\exp(-i\beta J_{y_1})\exp(-i\alpha J_z). \quad (2.12)$$

A little thought or alternatively, a direct reduction of equation (2.12) expressing J_{y_1} and $J_{z'}$ in terms of J_y and J_z shows that the same rotation (α, β, γ) will be produced by making first a rotation through an angle γ about the *original* z-axis, then a rotation through an angle β about the *original* y-axis and eventually a rotation through an angle α about the *original* z-axis. Again from equation (1.7) the rotation operator $D(\alpha, \beta, \gamma)$ is

$$D(\alpha, \beta, \gamma) = \exp(-i\alpha J_z)\exp(-i\beta J_y)\exp(-i\gamma J_z). \quad (2.13)$$

Often we shall use $R = (\alpha, \beta, \gamma)$ as an abreviation for the Euler angles specifying a rotation. If R_1 and R_2 refer to two successive rotations R_2R_1 is the combined rotation. The inverse rotation to R is $R^{-1} = (-\gamma, -\beta, -\alpha)$. With this notation we have

$$\begin{aligned} D(R_1)D(R_2) &= D(R_2R_1), \\ D^{-1}(R) &= D(R^{-1}). \end{aligned} \tag{2.14}$$

In an irreducible representation of the rotation group of dimension $2I+1$ corresponding to an angular momentum I the rotation $(\alpha\,\beta\,\gamma)$ is represented by the matrix‡

$$\langle IM'|D(\alpha\,\beta\,\gamma)|IM\rangle = \mathscr{D}^I_{M'M}(\alpha\,\beta\,\gamma). \tag{2.15}$$

The operator D^+ is the adjoint of D, hence its matrix elements are related to the matrix elements of D by

$$\begin{aligned} \langle IM|D^+|IM'\rangle &= \langle IM'|D|IM\rangle^* \\ &= (\mathscr{D}^I_{M'M})^*. \end{aligned}$$

Also because D is a unitary operator:

$$D^+(\alpha\,\beta\,\gamma) = D^{-1}(\alpha\,\beta\,\gamma) = D(-\gamma\; -\beta\; -\alpha),$$

hence

$$(\mathscr{D}^I_{MM'}(\alpha\,\beta\,\gamma))^* = \mathscr{D}^I_{M'M}(-\gamma\; -\beta\; -\alpha).$$

The operator D is unitary

$$D^+D = DD^+ = 1,$$

hence the rotation matrices 2.15 are unitary matrices, and

$$\begin{aligned} \sum_{M'} (\mathscr{D}^I_{M'N}(R))^* \mathscr{D}^I_{M'M}(R) &= \delta(M, N), \\ \sum_{M'} \mathscr{D}^I_{MM'}(R)(\mathscr{D}^I_{NM'}(R))^* &= \delta(M, N). \end{aligned} \tag{2.16}$$

‡ We use the convention of Rose [54], Messiah [46], that $D(\alpha\,\beta\,\gamma)$ rotates the system through Euler angles $(\alpha\,\beta\,\gamma)$, while others, (Wigner [78], Fano and Racah [31], Edmonds [22] and Rose [53]) use the opposite convention, i.e. that $D(\alpha\,\beta\,\gamma)$ rotates the system through angles $(-\alpha\; -\beta\; -\gamma)$. The relations between our rotation matrices and theirs are thus

$D(\alpha\,\beta\,\gamma)$	corresponds to $D(-\alpha\; -\beta\; -\gamma)$
$\mathscr{D}^I_{MN}(\alpha\,\beta\,\gamma)$	corresponds to $\mathscr{D}^I_{MN}(-\alpha\; -\beta\; -\gamma) = (-1)^{M-N}(\mathscr{D}^I_{MN}(\alpha\,\beta\,\gamma))^*$
$d^I_{MN}(\beta)$	corresponds to $d^I_{MN}(-\beta) = (-1)^{M-N}\, d^I_{MN}(\beta)$.

Because the basis vectors of the representation are chosen as eigenfunctions of J_z and D has the form of equation (2.13) the matrices simplify as follows:

$$\begin{aligned}\mathscr{D}^I_{MN}(\alpha\,\beta\,\gamma) &= \langle IM|\exp(-i\alpha J_z)\exp(-i\beta J_y)\exp(-i\gamma J_z)|IN\rangle,\\ &= e^{-i(\alpha M+\gamma N)}\langle IM|\exp(-i\beta J_y)|IN\rangle,\\ &= e^{-i(\alpha M+\gamma N)}\, d^I_{MN}(\beta). \qquad (2.17)\end{aligned}$$

Phases of the rotation matrices depend upon the convention adopted for the Euler angles and on the choice of phases of the matrix elements of **J**. With the Condon and Shortley choice of phases (equation (2.4)) the reduced rotation matrices d^I_{MN} are real and can be expressed explicitly as

$$d^j_{mn}(\beta) = \sum_t (-1)^t \frac{[(j+m)!\,(j-m)!\,(j+n)!\,(j-n)!]^{\frac{1}{2}}}{(j+m-t)!\,(j-n-t)!\,t!\,(t+n-m)!}\times$$
$$\times(\cos\beta/2)^{2j+m-n-2t}\,(\sin\beta/2)^{2t+n-m},$$

where the sum is taken over all values of t which lead to non-negative factorials. (Formulae for $j = \frac{1}{2}, 1, \frac{3}{2}, 2$; Table 1.)

In particular

$$\begin{aligned}d^I_{MN}(\pi) &= (-1)^{I+M}\delta(M,-N),\\ d^I_{MN}(2\pi) &= (-1)^{2I}\delta(M,N).\end{aligned} \qquad (2.18)$$

Symmetry relations for the matrices d^I_{MN} are listed in Appendix V. The rotation matrices reduce to spherical harmonics when M or $N = 0$

$$\mathscr{D}^I_{MO}(\alpha\,\beta\,\gamma) = (C_{IM}\,(\beta\,\alpha))^*. \qquad (2.19)$$

The second of the equations (2.18) gives

$$d^I_{MM'}(2\pi) = (-1)^{2I}\, d^I_{MM'}(0),$$

thus if I is half-integral the rotation matrix for $\beta = 2\pi$ has the opposite sign to the matrix for $\beta = 0$ and the rotation matrix is periodic in β with period 4π. Alternatively in the range $(0, 2\pi)$ the rotation matrices are double valued, the two values differing in phase by a factor of (-1). In terms of wave functions, a wave function corresponding to half-integral angular momentum changes sign on rotation through an angle 2π

about any axis. This arbitrariness in the sign of the wave functions leads to no arbitrariness of observable quantities provided the initial choice of phase is used consistently throughout the calculation. Care should be taken to use the same phase for identical rotations. Wigner [78] gives a detailed discussion of the double valuedness of the half-integral representations.

Besides unitarity the rotation matrices obey another orthogonality condition arising from a theorem of products of representations under group integration [Weyl [75]]. The theorem states that the products of matrix elements belonging to inequivalent representations of a group, and products of different elements of the same representation vanish on summation over all group elements (integration for a continuous group). Applied to the rotation group the theorem gives

$$\int_0^{2\pi}\int_0^{2\pi}\int_0^{\pi}(\mathscr{D}^I_{MM'}(\alpha\,\beta\,\gamma))^*\;\mathscr{D}^J_{NN'}(\alpha\,\beta\,\gamma)\sin\beta\,d\beta\,d\alpha\,d\gamma$$
$$=\frac{8\pi^2}{2I+1}\,\delta(M,N)\,\delta(M',N')\,\delta(I,J).$$

The normalization factor $8\pi^2/2I+1$ arises from the unitarity equation (2.16) of the matrices of the representation.

The particular case of $J = \frac{1}{2}$ deserves special mention because of its importance in discussion of spin. The components of $\mathbf{J}$ in the $J = \frac{1}{2}$ representation are represented by 2×2 matrices conveniently expressed in terms of the set of Pauli spin matrices $\boldsymbol{\sigma}$ by $\mathbf{J} = \frac{1}{2}\boldsymbol{\sigma}$.

If J_z is chosen to be diagonal and the choice of phases is made as in equation (2.4), then

$$\sigma_x = \begin{pmatrix}0 & 1\\ 1 & 0\end{pmatrix},\quad \sigma_y = \begin{pmatrix}0 & -i\\ i & 0\end{pmatrix},\quad \sigma_z = \begin{pmatrix}1 & 0\\ 0 & -1\end{pmatrix}.$$

The matrices $\boldsymbol{\sigma}$ have the anticommutation properties

$$\sigma_i\sigma_j+\sigma_j\sigma_i = 2\delta_{ij}$$

and in particular $\sigma_x^2 = \sigma_y^2 = \sigma_z^2 = \mathbf{1}$. Together with the 2×2 unit matrix $\mathbf{1}$ they are sufficient for a complete description of a $J = \frac{1}{2}$ system.

TABLE 1

Formulae for $d^j_{mm'}(\beta)$ *for* $j = \frac{1}{2}$, 1, $\frac{3}{2}$ *and* 2

$$d^{\frac{1}{2}}_{\frac{1}{2}\frac{1}{2}} = d^{\frac{1}{2}}_{-\frac{1}{2}-\frac{1}{2}} = \cos\left(\frac{\beta}{2}\right)$$

$$d^{\frac{1}{2}}_{-\frac{1}{2}\frac{1}{2}} = -d^{\frac{1}{2}}_{\frac{1}{2}-\frac{1}{2}} = \sin\left(\frac{\beta}{2}\right)$$

$$d^{1}_{11} = d^{1}_{-1-1} = \cos^2\left(\frac{\beta}{2}\right)$$

$$d^{1}_{1-1} = d^{1}_{-11} = \sin^2\left(\frac{\beta}{2}\right)$$

$$\begin{aligned} d^{1}_{01} &= d^{1}_{-10} = -d^{1}_{0-1} \\ &= -d^{1}_{10} = \sin\beta/\sqrt{2} \end{aligned}$$

$$d^{1}_{00} = \cos\beta$$

$$d^{\frac{3}{2}}_{\frac{3}{2}\frac{3}{2}} = d^{\frac{3}{2}}_{-\frac{3}{2}-\frac{3}{2}} = \cos^3\left(\frac{\beta}{2}\right)$$

$$\begin{aligned} d^{\frac{3}{2}}_{\frac{3}{2}\frac{1}{2}} &= d^{\frac{3}{2}}_{-\frac{1}{2}-\frac{3}{2}} = -d^{\frac{3}{2}}_{\frac{1}{2}\frac{3}{2}} \\ &= -d^{\frac{3}{2}}_{-\frac{3}{2}-\frac{1}{2}} \\ &= -\sqrt{3}\cos^2\left(\frac{\beta}{2}\right)\sin\left(\frac{\beta}{2}\right) \end{aligned}$$

$$\begin{aligned} d^{\frac{3}{2}}_{\frac{3}{2}-\frac{1}{2}} &= d^{\frac{3}{2}}_{-\frac{1}{2}\frac{3}{2}} = d^{\frac{3}{2}}_{\frac{1}{2}-\frac{3}{2}} \\ &= d^{\frac{3}{2}}_{-\frac{3}{2}\frac{1}{2}} = \sqrt{3}\cos\left(\frac{\beta}{2}\right)\sin^2\left(\frac{\beta}{2}\right) \end{aligned}$$

$$d^{\frac{3}{2}}_{\frac{3}{2}-\frac{3}{2}} = -d^{\frac{3}{2}}_{-\frac{3}{2}\frac{3}{2}} = -\sin^3\left(\frac{\beta}{2}\right)$$

$$\begin{aligned} d^{\frac{3}{2}}_{\frac{1}{2}\frac{1}{2}} &= d^{\frac{3}{2}}_{-\frac{1}{2}-\frac{1}{2}} \\ &= \cos\left(\frac{\beta}{2}\right)\left(3\cos^2\left(\frac{\beta}{2}\right)-2\right) \end{aligned}$$

$$\begin{aligned} d_{\frac{1}{2}-\frac{1}{2}}^{\frac{3}{2}} &= -d^{\frac{3}{2}}_{-\frac{1}{2}\frac{1}{2}} \\ &= \sin\left(\frac{\beta}{2}\right)\left(3\sin^2\left(\frac{\beta}{2}\right)-2\right) \end{aligned}$$

$$d^{2}_{22} = d^{2}_{-2-2} = \cos^4\left(\frac{\beta}{2}\right)$$

$$\begin{aligned} d^{2}_{21} &= -d^{2}_{12} = -d^{2}_{-2-1} \\ &= d^{2}_{-1-2} = -\tfrac{1}{2}\sin\beta(1+\cos\beta) \end{aligned}$$

$$\begin{aligned} d^{2}_{20} &= d^{2}_{02} = d^{2}_{-20} \\ &= d^{2}_{0-2} = \sqrt{\tfrac{3}{8}}\sin^2\beta \end{aligned}$$

$$\begin{aligned} d^{2}_{2-1} &= d^{2}_{1-2} = -d^{2}_{-21} \\ &= -d^{2}_{-12} = \tfrac{1}{2}\sin\beta(\cos\beta-1) \end{aligned}$$

$$d^{2}_{2-2} = d^{2}_{-22} = \sin^4\left(\frac{\beta}{2}\right)$$

$$\begin{aligned} d^{2}_{11} &= d^{2}_{-1-1} \\ &= \tfrac{1}{2}(2\cos\beta-1)(\cos\beta+1) \end{aligned}$$

$$\begin{aligned} d^{2}_{1-1} &= d^{2}_{-11} \\ &= \tfrac{1}{2}(2\cos\beta+1)(1-\cos\beta) \end{aligned}$$

$$\begin{aligned} d^{2}_{10} &= d^{2}_{0-1} = -d^{2}_{01} \\ &= -d^{2}_{-10} = -\sqrt{\tfrac{3}{2}}\sin\beta\cos\beta \end{aligned}$$

$$d^{2}_{00} = \tfrac{1}{2}(3\cos^2\beta-1)$$

References for additional tables for $d^j_{mm'}(\beta)$ are as follows:

j = 2, 4, 6: BUCKMASTER, H. A. (1964) *Can. J. Phys.* **42**, 386

j = 1, 3, 5: —— (1966) *Can. J. Phys.* **44**, 2525.

j = 3: YING-NAN CHIU (1966) *J. chem. Phys.* **45**, 2969.

The rotation matrices take a particularly simple form because $J_y = \frac{1}{2}\sigma_y$ and $\sigma_y^2 = 1$. Thus

$$\exp(-i\beta J_y) = \exp(-i\beta\sigma_y/2) = \mathbf{1}\cos\beta/2 - i\sigma_y \sin\beta/2.$$

Substitution of the explicit form of the matrices σ_y and $\mathbf{1}$ gives

$$d^{\frac{1}{2}}_{MM'}(\beta) = \begin{pmatrix} \cos\beta/2 & -\sin\beta/2 \\ \sin\beta/2 & \cos\beta/2 \end{pmatrix}.$$

2.5. Rotation Matrices as Symmetric Top Eigenfunctions

The rotation matrices $\mathscr{D}^I_{MN}(\alpha, \beta, \gamma)$ are eigenfunctions of the total angular momentum of a rigid body whose orientation is specified by the Euler angles α, β, γ. (These angles measure the orientation of the principal axes (x', y', z') fixed in the body relative to a set of axes (x, y, z) fixed in space.) The rotation matrices are also eigenfunctions of L_z and $L_{z'}$ with eigenvalues M and N respectively. If the rigid body has an axis of symmetry and the z'-body fixed axis is oriented in the direction of this axis then $L_{z'}$ as well as L_z are constants of the motion, and the rotation matrices are eigenfunctions of the Hamiltonian of the rigid rotator. These facts follow simply from the rotational properties of the wave function. Suppose $\phi(R)$ where $R = (\alpha\ \beta\ \gamma)$ is a wave function of the rigid rotator. A rotation of $\phi(R)$ by $R_1 = (\alpha_1\ \beta_1\ \gamma_1)$ produces a new wave function $\phi'(R) = D(R_1)\phi(R)$ and the value of the rotated wave function $\phi'(R)$ at the point R is the same as that of the old wave function at the point R' which is carried into R by the rotation R_1, i.e.

$$\phi'(R) = D(R_1)\phi(R) = \phi(R'). \tag{2.20}$$

If $\phi(R) = \phi_{IN}(R)$ is an eigenfunction of $\mathbf{L}^2$, L_z and H the Hamiltonian with eigenvalues $I(I+1)$, N and E, then $\phi'(R)$ must also be an eigenfunction of $\mathbf{L}^2$ and H with the same eigenvalues. If the eigenvalue E has only rotational degeneracy the state $\phi'(R)$ can be expanded in the set $\phi_{IM}(R)$. Thus

$$\begin{aligned}\phi'(R) = \phi_{IN}(R') &= \sum_M \phi_{IM}(R)\langle IM|D(R_1)|IN\rangle, \\ &= \sum_M \phi_{IM}(R)\,\mathscr{D}^I_{MN}(R_1).\end{aligned}$$

In this expression the ket $|IN\rangle$ in the matrix element stands for the state $\phi_{IN}(R)$. The relation takes an interesting form if $R_1 = R = (\alpha\,\beta\,\gamma)$, because then $R' = (0\ 0\ 0) = (0)$. Using the unitary property of the rotation matrices (2.16) we obtain

$$\phi_{IM}(R) = \sum_N (\mathscr{D}^I_{MN}(R))^* \phi_{IN}(0). \tag{2.21}$$

Thus the $2I+1$ wave functions $\phi_{IM}(R)$ are determined for all values of R by their value at $R = (0\ 0\ 0)$ and the rotational invariance of the Hamiltonian. Equation (2.21) gives the general form of the wave function of an asymmetric rigid rotator, [42], [80]. When the rotator has an axis of symmetry (chosen to be the z'-axis) there is a further specialization. An arbitrary rotation γ_0 about the symmetry axis z' should leave the wave function invariant up to a phase and $L_{z'}$ is conserved. This is possible if only one of the $\phi_{IN}(0)$ is non-zero and the wave function is $(\mathscr{D}^I_{MN}(R))^*$ apart from a normalization factor. The quantum numbers M and N are eigenvalues of L_z and $L_{z'}$ respectively.‡

Relation (2.20) may also be used to obtain explicit expressions for the components of **L** as differential operators. For example if R_1 is an infinitesimal rotation through an angle ϵ_z about the z-axis then $R' = (\alpha-\epsilon_z,\ \beta,\ \gamma)$ if $R = (\alpha,\ \beta,\ \gamma)$ and $D(R_1) = (1-i\epsilon_z L_z)$. Thus (2.20) becomes

$$(1-i\epsilon_z L_z)\phi(\alpha,\beta,\gamma) \approx \phi(\alpha-\epsilon_z,\beta,\gamma),$$

$$\approx \phi(\alpha,\beta,\gamma)-\epsilon_z\frac{\partial\phi}{\partial\alpha},$$

and
$$L_z = -i\frac{\partial}{\partial\alpha}.$$

Similar calculations using infinitesimal rotations about the x, y, and z'-axes give§

$$L_{\pm} = -ie^{\pm i\alpha}\left[-\cot\beta\frac{\partial}{\partial\alpha} \pm i\frac{\partial}{\partial\beta} + \frac{1}{\sin\beta}\frac{\partial}{\partial\gamma}\right], \qquad L_{z'} = -i\frac{\partial}{\partial\gamma},$$

‡ Bohr and Mottelson [13] use wave functions for the rigid rotator which are the complex conjugate of ours.

§ To obtain L_x and L_y we need expressions for $R' = (\alpha+d\alpha,\ \beta+d\beta,\ \gamma+d\gamma)$ in terms of the infinitesimal angles of rotation about the x, y, and z-axes.

while the expressions for L_z, $L_\pm$ give

$$\mathbf{L}^2 = \left[-\frac{\partial^2}{\partial\beta^2} - \cot\beta\frac{\partial}{\partial\beta} - \frac{1}{\sin^2\beta}\left(\frac{\partial^2}{\partial\alpha^2}+\frac{\partial^2}{\partial\gamma^2}\right)+\frac{2\cos\beta}{\sin^2\beta}\frac{\partial^2}{\partial\alpha\,\partial\gamma}\right].$$

The above discussion also holds for half integral angular momenta; but then the rotation matrices lose their meaning as eigenfunctions of a classical rigid rotator. They are, however, the approximate eigenfunctions of a particle of half integral spin coupled to a rigid rotator and in this form occur as the collective eigenfunctions of a deformed nucleus with odd atomic weight [13] [80] and of a molecule where the component of the electron angular momentum along the molecular symmetry axis is not zero.

2.6. The Vector Model and Classical Limits

As discussed in section 2.2 a representation of the rotation group is unique only up to a choice of basis for the manifold determining the representation. The basis vectors are chosen as eigenfunctions of the square of the total angular momentum $\mathbf{J}^2$ and its z-component J_z. There are, however, an infinite number of equivalent ways of choosing J_z corresponding to all possible directions of the z-axis. If for example, we are applying a perturbation which has axial symmetry about some direction it is most convenient to take J_z referred to this direction. Then the perturbing operator commutes with J_z and the perturbed states remain diagonal in J_z.

If a set of axes (x', y', z') is obtained by a rotation R from a set (x, y, z) then the eigenstates $|JN\rangle'$ of $J_{z'}$ are given by rotating the corresponding eigenstates $|JN\rangle$ of J_z with the

---

These may be obtained from expressions for the components of angular velocity of a rigid body and are

$$d\alpha = \epsilon_x \cot\beta\cos\alpha + \epsilon_y\cot\beta\sin\alpha - \epsilon_z$$

$$d\beta = \epsilon_x\sin\alpha - \epsilon_y\cos\alpha$$

$$d\gamma = -\epsilon_x\frac{\cos\alpha}{\sin\beta} - \epsilon_y\frac{\sin\alpha}{\sin\beta}.$$

axes, i.e. by transforming the state $|JN\rangle$ with the operator $D(R)$,

$$|JN\rangle' = D(R)|JN\rangle$$
$$= \sum_M |JM\rangle\langle JM|D(R)|JN\rangle$$
$$= \sum_M |JM\rangle \mathscr{D}^J_{MN}(R). \tag{2.22}$$

Equation (2.22) specifies the eigenstates of $J_{z'}$ in terms of the eigenstates of J_z. The states $\langle JN|$ conjugate to those in (2.22), rotate as

$$'\langle JN| = \langle JN|D^+ = \sum_M (\mathscr{D}^J_{MN})^* \langle JM|, \tag{2.23}$$

and we say they transform contragrediently (cogredience is defined by (2.22)). A symmetry property of rotation matrices (Appendix V)

$$\mathscr{D}^J_{MN}(R)^* = (-1)^{M-N} \mathscr{D}^J_{-M-N}(R),$$

shows that the transformation (2.23) for $\langle JN|$ is the same as that for $(-1)^N|J-N\rangle$; that is these two quantities behave in the same way under coordinate rotations.

Spherical harmonics afford a particular example of equation (2.22)

$$C_{ln}(\theta', \varphi') = \sum_m \mathscr{D}^l_{mn}(\alpha, \beta, \gamma) C_{lm}(\theta, \varphi). \tag{2.24}$$

The angles (θ, ϕ) and (θ', φ') are the angular coordinates of a point in the old and new coordinate systems. If $n = 0$ we obtain an addition theorem for spherical harmonics (cf. also section 4.6)

$$P_l(\cos\theta') = \sum_m C_{lm}(\beta, \alpha)^* C_{lm}(\theta, \varphi) \tag{2.25}$$

using equations (2.19) and (2.10).

Equation (2.22) leads to a geometrical interpretation of the rotation matrices. If we have a state with $\mathbf{J}^2 = J(J+1)$ and $J_z = M$ the indeterminacy of J_x and J_y is represented on the vector model by a vector $\mathbf{J}$ (of length $\sqrt{\{J(J+1)\}}$) precessing about O_z. If we make a measurement of the projection of $\mathbf{J}$ on another axis $O_{z'}$ inclined at an angle β to O_z (Fig. 3) the probability for finding a value M' is just $|\mathscr{D}^J_{M'M}(\alpha\,\beta\,\gamma)|^2$. On the vector model we should expect to find the spread of values $M_1 \leqslant M' \leqslant M_2$ shown in Fig. 3 due to the precession of $\mathbf{J}$

about O_z, with a probability $P(M')$ of finding a value M' assuming that **J** precesses uniformly about the O_z-axis,

$$P(M') = \frac{1}{\pi}[\mathbf{J}^2(1-\cos^2\beta)-(M^2+M'-2MM'\cos\beta)]^{-\frac{1}{2}}. \quad (2.26)$$

The effect of quantum indeterminacy is to allow values of M' outside the limits (M_1, M_2) given by the vector model, but

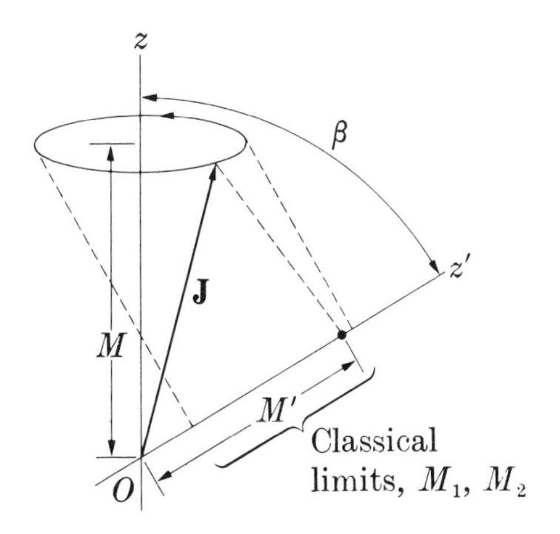

FIG. 3. Rotation of the quantization axis from O_z to $O_{z'}$. The classical limit indicates the spread of values for M' to be expected from a classical precession about O_z.

$|\mathscr{D}^J_{M'M}|^2$ falls off exponentially with M' in this region. Also $\mathscr{D}^J_{M'M}$ oscillates within the limits (M_1, M_2); but for large J, i.e. in the classical limit, $|\mathscr{D}^J_{M'M}|^2$ averaged over several values of M' to remove the oscillation is approximated by $P(M')$ obtained from the vector model. The asymptotic behaviour of the rotation matrices for large J has been discussed in detail in the W.K.B approximation by Brussaard and Tolhoek [15].

The general rotation includes the two Euler angles α and γ (section 2.4) for azimuthal rotation about the old and new z-axis respectively. These play the rôle of phase angles only, occurring as a factor $\exp -i(M\alpha+N\gamma)$. Since

$$|\mathscr{D}^J_{MN}(\alpha\,\beta\,\gamma)|^2 = |d^J_{MN}(\beta)|^2$$

they do not affect the probability interpretation just described.

2.7. Coupling of Two Angular Momenta

Often we work with systems made up of two or more parts, each with angular momenta. These may be different particles, or perhaps the spin and orbital properties of one particle. For the present we consider a system in which the total angular momentum $\mathbf{J}$ is the sum of components $\mathbf{J}_1$ and $\mathbf{J}_2$. If an interaction between the two parts is such as to leave the individual angular momenta and their z-components constants of the motion, a complete set of commuting operators would include H, $\mathbf{J}_1^2$, J_{1z}, $\mathbf{J}_2^2$ and J_{2z}. The corresponding eigenfunctions $|\alpha j_1 j_2 m_1 m_2\rangle$ may always be written in the simple product form

$$|\alpha j_1 j_2 m_1 m_2\rangle = \sum_{\beta\gamma} |\beta j_1 m_1\rangle|\gamma j_2 m_2\rangle,$$

where α, β, and γ represent any other quantum numbers needed to specify the states. In the following they will not be written explicitly. We have the eigenvalue equations

$$\mathbf{J}_1^2|j_1 j_2 m_1 m_2\rangle = j_1(j_1+1)|j_1 j_2 m_1 m_2\rangle,$$
$$J_{1z}|j_1 j_2 m_1 m_2\rangle = m_1|j_1 j_2 m_1 m_2\rangle,$$

for $\mathbf{J}_1^2$ and J_{1z} and similarly for $\mathbf{J}_2^2$ and J_{2z}. We could, however, choose a set including H, $\mathbf{J}_1^2$, $\mathbf{J}_2^2$, $\mathbf{J}^2 = (\mathbf{J}_1+\mathbf{J}_2)^2$ and $J_z = J_{1z}+J_{2z}$, which contains as many physical observables as before. The eigenfunctions $|j_1 j_2 JM\rangle$ now satisfy

$$\mathbf{J}^2|j_1 j_2 JM\rangle = J(J+1)|j_1 j_2 JM\rangle,$$
$$J_z|j_1 j_2 JM\rangle = M|j_1 j_2 JM\rangle,$$

while $\mathbf{J}_1^2$ and $\mathbf{J}_2^2$ have the same eigenvalues as before. In physical applications this is often a more useful set. For instance when an interaction between the two parts of the system is introduced as a perturbation $\mathbf{J}^2$ and $\mathbf{J}_z$ may be conserved, but not the individual z-components J_{1z} and J_{2z}.

The unitary transformation connecting these two representations

$$\begin{aligned} |j_1 j_2 JM\rangle &= \sum_{m_1 m_2} |j_1 j_2 m_1 m_2\rangle\langle j_1 j_2 m_1 m_2|j_1 j_2 JM\rangle, \\ |j_1 j_2 m_1 m_2\rangle &= \sum_{JM} |j_1 j_2 JM\rangle\langle j_1 j_2 JM|j_1 j_2 m_1 m_2\rangle, \end{aligned} \tag{2.27}$$

defines the vector-addition coefficient $\langle j_1 j_2 m_1 m_2 | j_1 j_2 JM \rangle = \langle j_1 j_2 JM | j_1 j_2 m_1 m_2 \rangle$ (sometimes called a Wigner or Clebsch–Gordan coefficient). Often for brevity we shall write the vector addition coefficient as $\langle j_1 j_2 m_1 m_2 | JM \rangle$ or even as $\langle m_1 m_2 | JM \rangle$ when confusion will not result. For given j_1 and j_2 the values of J are restricted by the 'triangular condition' [Dirac [20]]

$$j_1 + j_2 \geqslant J \geqslant |j_1 - j_2|,$$

and J ranges from $j_1 + j_2$ down to $|j_1 - j_2|$ in integer steps. Classically $\mathbf{J}$ is the sum of $\mathbf{j}_1$ and $\mathbf{j}_2$, so the magnitude of the vectors must be such that they can form three sides of a triangle. The triangle condition is symmetric in $j_1 j_2$ and J as suggested by the classical vector picture. Since $J_z = J_{1z} + J_{2z}$ the vector addition coefficient vanishes unless $M = m_1 + m_2$. The orthonormality of the eigenfunctions $|JM\rangle$ and $|j_1 j_2 m_1 m_2\rangle$ leads to the orthogonality relations for the coefficients

$$\sum_{m_1 m_2} \langle JM | j_1 j_2 m_1 m_2 \rangle \langle j_1 j_2 m_1 m_2 | J'M' \rangle = \delta(J, J')\, \delta(M, M'),$$

and

$$\sum_{JM} \langle j_1 j_2 m_1 m_2 | JM \rangle \langle JM | j_1 j_2 m_1' m_2' \rangle = \delta(m_1, m_1')\, \delta(m_2, m_2'), \quad (2.28)$$

which express the unitary nature of the transformation (2.27). Since each coefficient vanishes unless $M = m_1 + m_2$ the sum over M is purely formal in the second orthogonality relation and in fact the sum is only over J.

From a dynamical point of view the transformations (2.27) describe the addition of angular momentum. There is, however a geometric or group theoretic interpretation. The wave functions $|j_1 j_2 m_1 m_2\rangle$ for a two-component system in angular momentum states j_1, j_2 span a $(2j_1+1)(2j_2+1)$ manifold. On rotation of the coordinate system the wave functions for the two components transform separately according to representations $\mathscr{D}_{j_1}$ and $\mathscr{D}_{j_2}$ of the rotation group and the composite states transform as equation (2.22)

$$|j_1 j_2 n_1 n_2\rangle' = \sum_{m_1 m_2} \mathscr{D}^{j_1}_{m_1 n_1}(R)\, \mathscr{D}^{j_2}_{m_2 n_2}(R) |j_1 j_2 m_1 m_2\rangle. \quad (2.29)$$

The basis states of the composite system transform according to a $(2j_1+1)(2j_2+1)$ dimensional representation of the rotation group denoted by $\mathscr{D}_{j_1}\times\mathscr{D}_{j_2}$. This representation is reducible and the unitary transformation of equation (2.27) reduces it to its irreducible components and the states $|JM\rangle$ are basis vectors of the reduced representation. It can be shown that J runs from j_1+j_2 down to $|j_1-j_2|$. We may write the reduction symbolically as

$$\mathscr{D}_{j_1}\times\mathscr{D}_{j_2} = \sum_{J=|j_1-j_2|}^{j_1+j_2} \mathscr{D}_J \tag{2.30}$$

Writing the reduction explicitly yields relations between rotation matrices. If D is an arbitrary rotation we have

$$\begin{aligned}\mathscr{D}^J_{MN} &= \langle JM|D|JN\rangle \\ &= \sum_{m_1 m_2 n_1 n_2} \langle JM|m_1 m_2\rangle\langle m_1 m_2|D|n_1 n_2\rangle\langle n_1 n_2|JN\rangle \\ &= \sum_{m_1 m_2 n_1 n_2} \langle JM|m_1 m_2\rangle\, \mathscr{D}^{j_1}_{m_1 n_1}\mathscr{D}^{j_2}_{m_2 n_2}\langle n_1 n_2|JN\rangle \end{aligned} \tag{2.31}$$

and similarly for the inverse relation

$$\mathscr{D}^{j_1}_{m_1 n_1}\mathscr{D}^{j_2}_{m_2 n_2} = \sum_{JMN} \langle m_1 m_2|JM\rangle\, \mathscr{D}^J_{MN}\langle JN|n_1 n_2\rangle. \tag{2.32}$$

2.7.1. The Vector Model

In terms of the vector model the state $|j_1 j_2 JM\rangle$ is represented by the two vectors $\mathbf{j}_1$ and $\mathbf{j}_2$ precessing in phase about their resultant $\mathbf{J}$ (which in turn precesses about the z-axis) (Fig. 4(a)). The precession of $\mathbf{j}_1$ and $\mathbf{j}_2$ about $\mathbf{J}$ and its projection on the z-axis represents the indeterminacy in their individual z-components m_1 and m_2 although their sum M remains constant. The square of the vector addition coefficient

$$|\langle j_1 j_2 m_1 m_2|JM\rangle|^2$$

is the probability that a measurement in the state $|JM\rangle$ gives the particular values m_1 and m_2 for J_{1z} and J_{2z}. Conversely in the state $|j_1 j_2 m_1 m_2\rangle$ the two vectors precess independently about the z-axis (Fig. 4(b)) and $|\langle JM|j_1 j_2 m_1 m_2\rangle|^2$ is the probability that at any instant their resultant will be J. As in the case of rotation matrices the vector model suggests an expression for the squares of the vector coupling coefficients in the limit of large quantum numbers. A comparison of Figs. 3 and 4(a) leads to expressions for the vector-

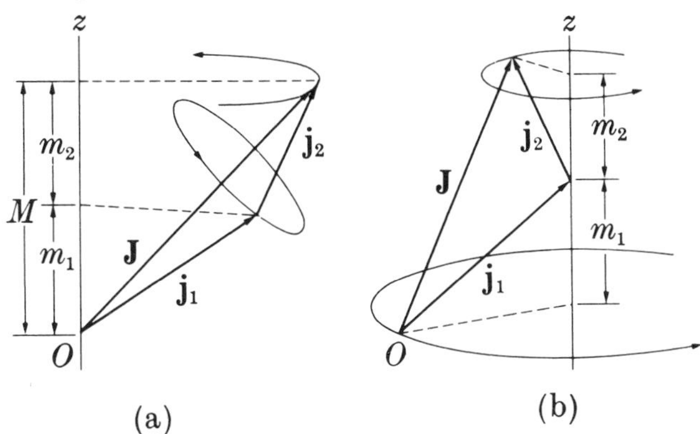

FIG. 4. Vector picture of two angular momenta coupled to give a resultant $\mathbf{J}$.
(a) In the (JM) representation $\mathbf{j}_1$ and $\mathbf{j}_2$ are coupled to give a resultant $\mathbf{J}$ which is precessing about O_z. m_1 and m_2 are undetermined.
(b) In the $(j_1m_1j_2m_2)$ representation $\mathbf{j}_1$ and $\mathbf{j}_2$ precess independently about O_z leading to uncertainty in $\mathbf{J}$.

coupling coefficients which should hold in the limit of large quantum numbers.

If $j_1 \gg j_2$, and hence $J \gg j_2$, then

$$\langle j_1 j_2 m_1 m_2 | JM \rangle \approx d^{j_2}_{m_2, J-j_1}(\beta)$$

and if $m_2 = 0$ and $j_1 = J$

$$\langle J j_2 M O | JM \rangle \approx P_{j_2}(\cos\beta)$$

where $\cos\beta = M/J$. These and other relations are derived by Brussard and Tolhoek [15]; see also [96] and Appendix VII.

2.7.2. *Explicit Formula for Vector Addition Coefficients*

Recurrence relations for the coefficients can be obtained from the operator identities $J_\pm = J_{1\pm} + J_{2\pm}$. In matrix form these become

$$\langle JM \pm 1 | J_\pm | JM \rangle$$
$$= \sum_{\substack{m_1 m_2 \\ n_1 n_2}} \langle JM \pm 1 | n_1 n_2 \rangle \langle n_1 n_2 | J_{1\pm} + J_{2\pm} | m_1 m_2 \rangle \langle m_1 m_2 | JM \rangle.$$

Matrix multiplication from the left by $\langle m_1' m_2' | JM' \rangle$, use of the orthogonality relation equation (2.27) and the matrix elements of $J_{\pm}$ (2.4), give us

$$\{(J\pm M+1)(J\mp M)\}^{\frac{1}{2}}\langle m_1 m_2|JM\pm 1\rangle$$
$$= \{(j_1\mp m_1+1)(j_1\pm m_1)\}^{\frac{1}{2}}\langle m_1\mp 1 m_2|JM\rangle+$$
$$+\{(j_2\mp m_2+1)(j_2\pm m_2)\}^{\frac{1}{2}}\langle m_1 m_2\mp 1|JM\rangle. \quad (2.33)$$

These relations are sufficient to determine the vector addition coefficients.

The left-hand side of equation (2.33) vanishes if we take the upper sign and put $M = J$. With the normalization condition (2.28) this enables us to determine the various $\langle m_1 m_2|JJ\rangle$, apart from an overall phase; this we fix (following Racah [48]) ‡ by the convention that $\langle j_1 J-j_1|JJ\rangle$ is always real and positive. The lower sign in equation (2.33) gives us $\langle m_1 m_2|JM-1\rangle$ in terms of $\langle m_1' m_2'|JM\rangle$, so by a 'ladder calculation' starting with $M = J$ we get all the coefficients, which, we see, must all be real. After some heavy algebra along these lines Racah obtained the general formula

$$\langle ab\alpha\beta|c\gamma\rangle = \delta(\alpha+\beta,\gamma)\Delta(a\,b\,c)\times$$
$$\times[(2c+1)(a+\alpha)!\,(a-\alpha)!\,(b+\beta)!\,(b-\beta)!\,(c+\gamma)!\,(c-\gamma)!]^{\frac{1}{2}}\times$$
$$\times\sum_{\nu}(-1)^{\nu}[(a-\alpha-\nu)!\,(c-b+\alpha+\nu)!\,(b+\beta-\nu)!\times$$
$$\times(c-a-\beta+\nu)!\,\nu!\,(a+b-c-\nu)!]^{-1},$$

where

$$\Delta(abc) = \left[\frac{(a+b-c)!\,(a+c-b)!\,(b+c-a)!}{(a+b+c+1)!}\right]^{\frac{1}{2}}, \quad (2.34)$$

and ν runs over all values which do not led to negative factorials.

The formula can be simplified when $\alpha = \beta = \gamma = 0$ (Racah [4]); if $2g = a+b+c$, $\langle ab00|c0\rangle = 0$ if $2g$ is odd, and

$$\langle ab00|c0\rangle = (-1)^{g+c}(2c+1)^{\frac{1}{2}}\,\Delta(abc)g!\,[(g-a)!\,(g-b)!\,(g-c)!]^{-1} \quad (2.35)$$

if $2g$ is even (cf. section 4.7.2 and Table 2)

‡ With identical results, Condon and Shortley [17] adopt the convention that $\langle j_1 j_2|j_1+j_2,\, j_1+j_2\rangle = +1$ and that the matrix elements

$$\langle j_1 j_2 JM|J_{1z}|j_1 j_2 J-1\;M\rangle$$

are real and positive for all J.

Obtaining numerical values from the general formula is tedious, but by assigning a definite value to one argument, say b, equation (2.34) reduces to simple closed forms. Table 3 contains formulae for the more symmetric Wigner 3-j symbol defined in equation (3.3) for $b = 0, \frac{1}{2}, 1, \frac{3}{2}, 2$. Symmetry relations and other formulae are given in Appendix I. Extensive numerical tables have been published [58] [66], and for $\langle ab\frac{1}{2}-\frac{1}{2}|c0\rangle$ by de Shalit [18]. Values for $\alpha = \beta = \gamma = 0$ are given in Table 2.

TABLE 2

Values of $\begin{pmatrix} a & b & c \\ 0 & 0 & 0 \end{pmatrix}^2$. *An asterisk means the symbol is negative.*

a	b	c		a	b	c		a	b	c	
0	1	1	1/3*	1	5	6	6/143	3	3	6	100/3003
0	2	2	1/5	2	2	2	2/35*	3	4	5	20/1001
0	3	3	1/7*	2	2	4	2/35	3	5	6	7/429*
0	4	4	1/9	2	3	3	4/105	4	4	4	18/1001
0	5	5	1/11*	2	3	5	10/231*	4	4	6	20/1287*
0	6	6	1/13	2	4	4	20/693*	4	5	5	2/143*
1	1	2	2/15	2	4	6	5/143	4	6	6	28/2431
1	2	3	3/35*	2	5	5	10/429	5	5	6	80/7293
1	3	4	4/63	2	6	6	14/715*	6	6	6	400/46189*
1	4	5	5/99*	3	3	4	2/77*				

2.7.3. *Exchange Symmetry of Two-particle States*

When eigenstates of total angular momentum of two *identical* particles are constructed according to (2.27), and the individual angular momenta are the same, $j_1 = j_2$, the symmetry of the states under exchange of the two particles is determined by the symmetry of the vector addition coefficients

$$\langle jjmm'|JM\rangle = (-1)^{J-2j}\langle jjm'm|JM\rangle.$$

If we denote the state occupied by the ith particle as $|\rangle_i$, the coupled state may be written

$$|jjJM\rangle = \sum_{mm'} |jm\rangle_1|jm'\rangle_2\langle jjmm'|JM\rangle,$$

$$= \tfrac{1}{2}\sum_{mm'} \{|jm\rangle_1|jm'\rangle_2+(-1)^{J-2j}|jm'\rangle_1|jm\rangle_2\}\langle jjmm'|JM\rangle.$$

TABLE 3

Algebraic formulae for $\begin{pmatrix} a & b & c \\ \alpha & \beta & \gamma \end{pmatrix}$ *with* $c = \frac{1}{2}$, 1, $\frac{3}{2}$ *and* 2.

$$\begin{pmatrix} a & a+\frac{1}{2} & \frac{1}{2} \\ \alpha & -\alpha-\frac{1}{2} & \frac{1}{2} \end{pmatrix} = (-)^{a-\alpha-1} \left[\frac{a+\alpha+1}{(2a+2)(2a+1)}\right]^{\frac{1}{2}}$$

$$\begin{pmatrix} a & a & 1 \\ \alpha & -\alpha-1 & 1 \end{pmatrix} = (-)^{a-\alpha} \left[\frac{(a-\alpha)(a+\alpha+1)}{2a(a+1)(2a+1)}\right]^{\frac{1}{2}}$$

$$\begin{pmatrix} a & a & 1 \\ \alpha & -\alpha & 0 \end{pmatrix} = (-)^{a-\alpha} \frac{\alpha}{[a(a+1)(2a+1)]^{\frac{1}{2}}}$$

$$\begin{pmatrix} a & a+1 & 1 \\ \alpha & -\alpha-1 & 1 \end{pmatrix} = (-)^{a-\alpha} \left[\frac{(a+\alpha+1)(a+\alpha+2)}{(2a+1)(2a+2)(2a+3)}\right]^{\frac{1}{2}}$$

$$\begin{pmatrix} a & a+1 & 1 \\ \alpha & -\alpha & 0 \end{pmatrix} = (-)^{a-\alpha-1} \left[\frac{(a-\alpha+1)(a+\alpha+1)}{(a+1)(2a+1)(2a+3)}\right]^{\frac{1}{2}}$$

$$\begin{pmatrix} a & a+\frac{1}{2} & \frac{3}{2} \\ \alpha & -\alpha-\frac{3}{2} & \frac{3}{2} \end{pmatrix} = (-)^{a-\alpha-1} \left[\frac{3(a+\alpha+1)(a+\alpha+2)(a-\alpha)}{2a(2a+1)(2a+2)(2a+3)}\right]^{\frac{1}{2}}$$

$$\begin{pmatrix} a & a+\frac{1}{2} & \frac{3}{2} \\ \alpha & -\alpha-\frac{1}{2} & \frac{1}{2} \end{pmatrix} = (-)^{a-\alpha} \left[\frac{a+\alpha+1}{2a(2a+1)(2a+2)(2a+3)}\right]^{\frac{1}{2}}(a-3\alpha)$$

$$\begin{pmatrix} a & a+\frac{3}{2} & \frac{3}{2} \\ \alpha & -\alpha-\frac{3}{2} & \frac{3}{2} \end{pmatrix} = (-)^{a-\alpha-1} \left[\frac{(a+\alpha+1)(a+\alpha+2)(a+\alpha+3)}{(2a+1)(2a+2)(2a+3)(2a+4)}\right]^{\frac{1}{2}}$$

$$\begin{pmatrix} a & a+\frac{3}{2} & \frac{3}{2} \\ \alpha & -\alpha-\frac{1}{2} & \frac{1}{2} \end{pmatrix} = (-)^{a-\alpha} \left[\frac{3(a-\alpha+1)(a+\alpha+1)(a+\alpha+2)}{(2a+1)(2a+2)(2a+3)(2a+4)}\right]^{\frac{1}{2}}$$

$$\begin{pmatrix} a & a & 2 \\ \alpha & -\alpha-2 & 2 \end{pmatrix} = (-)^{a-\alpha} \left[\frac{3(a+\alpha+1)(a+\alpha+2)(a-\alpha-1)(a-\alpha)}{a(2a+3)(2a+2)(2a+1)(2a-1)}\right]^{\frac{1}{2}}$$

$$\begin{pmatrix} a & a & 2 \\ \alpha & -\alpha-1 & 1 \end{pmatrix} = (-)^{a-\alpha} \left[\frac{3(a-\alpha)(a+\alpha+1)}{a(2a+3)(2a+2)(2a+1)(2a-1)}\right]^{\frac{1}{2}}(2\alpha+1)$$

$$\begin{pmatrix} a & a & 2 \\ \alpha & -\alpha & 0 \end{pmatrix} = (-)^{a-\alpha} \frac{3\alpha^2-a(a+1)}{[a(a+1)(2a+3)(2a+1)(2a-1)]^{\frac{1}{2}}}$$

$$\begin{pmatrix} a & a+1 & 2 \\ \alpha & -\alpha-2 & 2 \end{pmatrix} = (-)^{a-\alpha} \left[\frac{(a+\alpha+1)(a+\alpha+2)(a+\alpha+3)(a-\alpha)}{a(a+1)(2a+4)(2a+3)(2a+1)}\right]^{\frac{1}{2}}$$

$$\begin{pmatrix} a & a+1 & 2 \\ \alpha & -\alpha-1 & 1 \end{pmatrix} = (-)^{a-\alpha-1} \left[\frac{(a+\alpha+2)(a+\alpha+1)}{a(a+1)(2a+4)(2a+3)(2a+1)}\right]^{\frac{1}{2}}(a-2\alpha)$$

$$\begin{pmatrix} a & a+1 & 2 \\ \alpha & -\alpha & 0 \end{pmatrix} = (-)^{a-\alpha-1}\, \alpha \left[\frac{3(a+\alpha+1)(a-\alpha+1)}{a(a+1)(a+2)(2a+3)(2a+1)}\right]^{\frac{1}{2}}$$

$$\begin{pmatrix} a & a+2 & 2 \\ \alpha & -\alpha-2 & 2 \end{pmatrix} = (-)^{a-\alpha} \left[\frac{(a+\alpha+1)(a+\alpha+2)(a+\alpha+3)(a+\alpha+4)}{(2a+5)(2a+4)(2a+3)(2a+2)(2a+1)}\right]^{\frac{1}{2}}$$

$$\begin{pmatrix} a & a+2 & 2 \\ \alpha & -\alpha-1 & 1 \end{pmatrix} = (-)^{a-\alpha-1} \left[\frac{(a+\alpha+1)(a+\alpha+2)(a+\alpha+3)(a-\alpha+1)}{(a+1)(a+2)(2a+1)(2a+3)(2a+5)}\right]^{\frac{1}{2}}$$

$$\begin{pmatrix} a & a+2 & 2 \\ \alpha & -\alpha & 0 \end{pmatrix} = (-)^{a-\alpha} \left[\frac{3(a+\alpha+1)(a+\alpha+2)(a-\alpha+1)(a-\alpha+2)}{(a+1)(2a+5)(2a+4)(2a+3)(2a+1)}\right]^{\frac{1}{2}}$$

Thus when particles 1 and 2 are interchanged it merely multiplies the state vector $|jjJm\rangle$ by $(-1)^{J-2j}$. That is, the state is symmetric or antisymmetric under exchange as $(J-2j)$ is even or odd. If j represents an orbital angular momentum, $2j$ is even, and the condition is whether the resultant J is even or odd. On the other hand if j is the total (spin plus orbit) for each particle ($j-j$ coupling), j will be integral or half-integral according to whether the particles are bosons or fermions. Thus both the symmetric boson states and the antisymmetric fermion states will have J even only, odd J states having the wrong symmetry in both cases.

A very simple example is given by the total spin states of two spin $-\frac{1}{2}$ fermions, $j = \frac{1}{2}$. The singlet $J = 0$ state is antisymmetric, the triplet $J = 1$ state is symmetric.

Of course, this simple property no longer holds when $j_1 \neq j_2$, and the exchange symmetry is no longer determined by the vector coupling.

CHAPTER III

COUPLING ANGULAR MOMENTUM VECTORS AND TRANSFORMATION THEORY

3.1. Transformation Theory

Often there are several degenerate but independent states $|\alpha\rangle_1$, $|\alpha\rangle_2$ ... which are eigenstates of some operator $\boldsymbol{\alpha}$ with the same eigenvalue α. Thus we require other labels to distinguish them. These may be provided by finding another operator $\boldsymbol{\beta}$ which commutes with $\boldsymbol{\alpha}$. Hence $\boldsymbol{\beta}$ has eigenstates which are linear superpositions of the $|\alpha\rangle_i$ belonging to the same eigenvalue α. It is then straightforward to diagonalize in this subspace to find these superpositions $|\alpha,\beta\rangle$ which are now labelled by α and the various β. If two or more combinations are still degenerate, i.e. share the same eigenvalues α and β, we need yet another operator $\boldsymbol{\gamma}$ which commutes with both $\boldsymbol{\alpha}$ and $\boldsymbol{\beta}$. So we proceed until we have resolved the original set of states $|\alpha\rangle_i$, degenerate in α, into a set completely non-degenerate in the eigenvalues α, β, γ, ... of a set of commuting operators $\boldsymbol{\alpha},\boldsymbol{\beta},\boldsymbol{\gamma}$, ... Such a set is called complete and represents the maximum number of simultaneously measured data allowed by the uncertainty principle.

The most important of these operators is usually the Hamiltonian H of the system or at least the principal part of it, the rest perhaps to be treated as a perturbation later. The other operators then have to be chosen to commute with H.

The need for the present chapter arises because there are often two or more ways of choosing our complete set. These sets of course are not independent but have eigenstates which are related by unitary linear transformation. If two such sets are denoted by A and B we may write

$$|A\rangle = \textstyle\sum_B |B\rangle\langle B|A\rangle. \tag{3.1}$$

The expansion coefficients $\langle B|A\rangle$ are the transformation amplitudes and form a unitary matrix. We have already met with two examples in sections 2.6 and 2.7. One is the transformation (2.22) linking states referred to differently oriented quantization axes, the other is the change of representation (2.27) for states comprising two angular momenta. Below we extend the latter example to systems made up of more than two angular momenta.

3.2. Scalar Contraction of Angular Momentum States

We have so far met the vector addition coefficient as giving a rotationally cogredient linear superposition of the products of two functions which separately behave cogrediently under rotations. Alternatively, it projects out the various irreducible parts of such products. A different point of view put forward by Wigner [78] considers the coupling of three angular momentum vectors to a zero resultant; and because the wave function of the coupled state has zero angular momentum it is independent of the choice of axes, i.e. it is a scalar or invariant quantity. Equation (2.27) and the knowledge that

$$\langle cc\gamma\,-\gamma|00\rangle = (-1)^{c-\gamma}/\sqrt{(2c+1)}$$

make it easy to perform, the scalar contraction product of three state vectors and we find that

$$\sum_{\alpha\beta\gamma} |a\alpha\rangle|b\beta\rangle|c\gamma\rangle\langle ab\alpha\beta|c-\gamma\rangle(-1)^{c-\gamma} \tag{3.2}$$

is an invariant.

We shall meet the contraction (3.2) in another guise in section 4.7 where we discuss the Wigner–Eckart theorem.

Among the various notations (listed in Appendix I) used for vector addition coefficients the Wigner $3-j$ symbol emphasises this contraction property. It is related to our present transformation coefficient by

$$\begin{pmatrix} a & b & c \\ \alpha & \beta & \gamma \end{pmatrix} \sqrt{(2c+1)} = (-1)^{a-b-\gamma}\langle ab\alpha\beta|c-\gamma\rangle, \tag{3.3}$$

provided, of course, that $\alpha+\beta+\gamma = 0$. Now the result 3.2 becomes

$$\sum_{\alpha\beta\gamma} |a\alpha\rangle|b\beta\rangle|c\gamma\rangle \begin{pmatrix} a & b & c \\ \alpha & \beta & \gamma \end{pmatrix} = \text{scalar invariant.} \tag{3.4}$$

In addition, use of the $3-j$ symbol often facilitates algebraic manipulation because of its high degree of symmetry. It is invariant under an even—cyclic—permutation in the order of its arguments, while an odd—non-cyclic—permutation merely multiplies it by $(-1)^{a+b+c}$. It thus avoids the unsymmetrical surds and phases appearing in the corresponding Clebsch-Gordan symmetry relations. The origin of these symmetries also appears more clearly through the contraction (3.4). All three angular momenta a, b, c, enter on an equal footing, thus the scalar invariant should be unchanged within a factor ± 1 by re-ordering them. A special case with $a = b = c = 1$ is the triple scalar product of vectors‡

$$\mathbf{u} \wedge \mathbf{v} \cdot \mathbf{w} = \mathbf{v} \wedge \mathbf{w} \cdot \mathbf{u} = \mathbf{w} \wedge \mathbf{u} \cdot \mathbf{v}$$
$$= -i\sqrt{6} \sum_{\lambda\mu\nu} u_\lambda v_\mu w_\nu \begin{pmatrix} 1 & 1 & 1 \\ \lambda & \mu & \nu \end{pmatrix}.$$

The $3-j$ symbol also has the visual advantage of displaying

‡ The quantities u_λ, v_μ, w_ν are the spherical components of the vectors **u**, **v**, and **w** defined in equation (4.10).

z-components on a different level from the vectors in accord with their inferior role.

Specializing equation (3.4) to the case $c = 0$ (and therefore $a = b$, $\alpha+\beta = 0$) allows us to define also a $1-j$ symbol, the analogue of the metric tensor. We have

$$\sum_{\alpha\beta} |a\alpha\rangle|b\beta\rangle\begin{pmatrix} & a \\ \alpha & \beta\end{pmatrix} = \text{a scalar},$$

where

$$\begin{pmatrix} & a \\ \alpha & \beta\end{pmatrix} = (-1)^{a+\alpha}\,\delta_{\alpha,-\beta}. \tag{3.5}$$

In many physical problems measured quantities are independent of the choice of coordinate axes, hence these quantities may be expressed in terms of scalars. In the course of a calculation of such a quantity we may be confronted with a complicated expression containing products of vector addition coefficients, and the solution of the problems often lies in extracting the various scalars. In the following sections we give some of the scalar invariants which are common to many problems.

3.3. Coupling of Three Angular Momenta

When we have three angular momentum vectors we may use an uncoupled representation

$$|j_1 m_1, j_2 m_2, j_3 m_3\rangle,$$

or one in which the vectors couple to a resultant J and M, that is, an eigenstate of $\mathbf{J}^2 = (\mathbf{j}_1+\mathbf{j}_2+\mathbf{j}_3)^2$ and $J_z = j_{1z}+j_{2z}+j_{3z}$. However, the latter is no longer unique, and we require a further quantum number. There are three possibilities: We may couple $\mathbf{j}_1$ and $\mathbf{j}_2$ to form $\mathbf{J}_{12}$, then add $\mathbf{j}_3$ vectorially to give $\mathbf{J}$. First

$$|j_1 j_2 J_{12} M_{12}\rangle = \sum_{m_1 m_2} |j_1 m_1\rangle|j_2 m_2\rangle\langle j_1 j_2 m_1 m_2|J_{12} M_{12}\rangle,$$

then

$$|(j_1 j_2)J_{12}, j_3; JM\rangle = \sum_{m_{12} m_3} |j_1 j_2 J_{12} M_{12}\rangle|j_3 m_2\rangle\langle J_{12} j_3 M_{12} m_3|JM\rangle. \tag{3.6}$$

This state is also an eigenfunction of $\mathbf{J}_{12}^2 = (\mathbf{j}_1+\mathbf{j}_2)^2$, and J_{12}

provides the additional quantum number which specifies the state of the coupled 3-vector system completely.

Alternatively, we may first combine $\mathbf{j}_2$ and $\mathbf{j}_3$ to give $\mathbf{J}_{23}$ and then add $\mathbf{j}_1$ to give $\mathbf{J}$, that is

$$|j_2 j_3 J_{23} M_{23}\rangle = \sum_{m_2 m_3} |j_2 m_2\rangle |j_3 m_3\rangle \langle j_2 j_3 m_2 m_3 | J_{23} M_{23}\rangle$$

then

$$|j_1, (j_2 j_3) J_{23}; JM\rangle = \sum_{m_1 M_{23}} |j_1 m_1\rangle |j_2 j_3 J_{23} M_{23}\rangle \langle j_1 J_{23}\, m_1 M_{23} | JM\rangle. \quad (3.7)$$

So we now have an eigenfunction of $\mathbf{J}_{23}^2 = (j_2 + j_3)^2$. Similarly we could have used $\mathbf{J}_{13}$, the resultant of $\mathbf{j}_1$ and $\mathbf{j}_3$, to define a set of states $|(j_1 j_3) J_{13}, j_2; JM\rangle$. Clearly, these three representations are not independent, and since they span the same subspace they must be connected by a linear transformation (equation (3.1)). For example

$$|(j_1 j_2) J_{12}, j_3; JM\rangle = \sum_{J_{23}} |j_1, (j_2 j_3) J_{23}; JM\rangle \times \\ \times \langle j_1, (j_2 j_3) J_{23}; J | (j_1 j_2) J_{12}, j_3; J\rangle \quad (3.8)$$

The transformation coefficient is a scalar and independent of M. We use the coefficient to define the Racah W-function [48],

$$\langle j_1, (j_2 j_3) J_{23}; J | (j_1 j_2) J_{12}, j_3; J\rangle = \\ = [(2J_{12}+1)(2J_{23}+1)]^{\frac{1}{2}} W(j_1 j_2 J j_3; J_{12} J_{23}), \quad (3.9)$$

whose normalization is chosen to simplify its symmetry properties. We may easily express it in terms of vector addition coefficients. We use (3.6) and (3.7) to expand both sides of (3.8) in the uncoupled representation $|j_1 m_1 j_2 m_2 j_3 m_3\rangle$. Equating coefficients we obtain

$$\langle j_1 j_2 m_1 m_2 | J_{12} M_{12}\rangle \langle J_{12}\; j_3 M_{12}\, m_3 | JM\rangle = \\ = \sum_{J_{23}} \langle j_2 j_3 m_2 m_3 | J_{23} M_{23}\rangle \langle j_1 J_{23}\, m_1 M_{23} | JM\rangle \times \\ \times \langle j_1, (j_2 j_3) J_{23}; J | (j_1 j_2)\, J_{12}, j_3; J\rangle. \quad (3.10)$$

For clarity, we rewrite this using a, b, c, d, e and f for the vectors and α, β, γ, δ, ε, and φ for the corresponding z-components

$$\langle ab\alpha\beta | e\alpha+\beta\rangle \langle ed\alpha+\beta, \gamma-\alpha-\beta | c\gamma\rangle \\ = \sum_f \langle bd\beta, \gamma-\alpha-\beta | f\gamma-\alpha\rangle \langle af\alpha, \gamma-\alpha | c\gamma\rangle \times \\ \times [(2e+1)(2f+1)]^{\frac{1}{2}}\, W(abcd; ef). \quad (3.11)$$

Multiplication of both sides by $\langle bd\ \beta, \gamma-\alpha-\beta|f'\gamma-\alpha\rangle$ and summing over β leads to the further relation

$$\langle af\alpha, \gamma-\alpha|c\gamma\rangle[2e+1)(2f+1)]^{\frac{1}{2}}W(abcd;\ ef) =$$
$$= \textstyle\sum_\beta \langle ab\alpha\beta|e\alpha+\beta\rangle\langle ed\alpha+\beta,\gamma-\alpha-\beta|c\gamma\rangle\langle bd\beta, \gamma-\alpha-\beta|f\gamma-\alpha\rangle. \quad (3.12)$$

In a similar manner we finally obtain

$$[(2e+1)(2f+1)]^{\frac{1}{2}}W(abcd;\ ef)\ \delta(cc')\ \delta(\gamma\gamma') =$$
$$= \textstyle\sum_{\alpha\beta} \langle ab\alpha\beta|e\alpha+\beta\rangle\langle ed\alpha+\beta,\ \gamma-\alpha-\beta|c\gamma\rangle\times \quad (3.13)$$
$$\times\langle bd\beta,\ \gamma-\alpha-\beta|f\gamma-\alpha\rangle\langle af\alpha\gamma-\alpha|c'\gamma'\rangle,$$

in the summation of which γ is held constant. In this last form the W-function appears not as transformation coefficient but as the scalar invariant obtained by contraction of four vector addition coefficients. In this, perhaps, lies its greatest value.

The symmetry properties of the W listed in Appendix II can be obtained easily from equation (3.13) by considering the corresponding symmetries of the vector addition coefficients involved. Equation (3.13) also makes it clear that the following triads of vectors have to satisfy 'triangular conditions' (section 2.7): (acf), (abe), (bdf), and (cde). In diagrams introduced by Levinson [85] lines represent angular momenta and triangular conditions must be obeyed at each vertex (cf. Chapter VII).

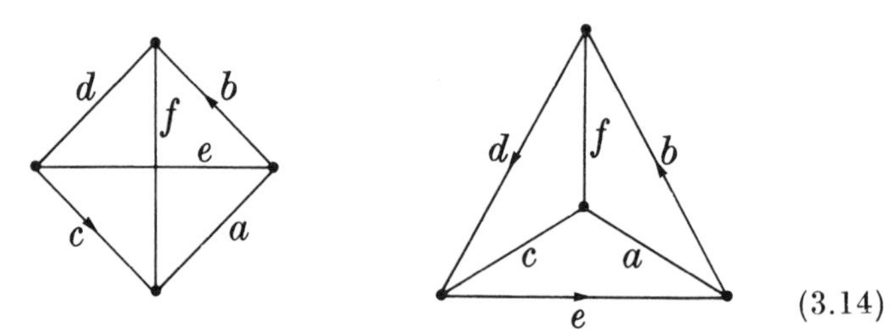

(3.14)

Equation (3.13) may be put into a form which shows the symmetry of the Racah W-function in a more obvious way by using the $3-j$ symbol defined in equation (3.3) (cf. Appendix II). This form is probably the most useful for calculations.

A general formula for the Racah functions may be found from (3.13) by using expression (2.34) for the vector addition coefficients. Racah [48] succeeded in reducing this to

$$W(abcd;\,ef) = \Delta(a\,b\,e)\,\Delta(a\,c\,f)\,\Delta(b\,d\,f)\,\Delta(c\,d\,e)\times$$
$$\times\sum_z(-1)^z(a+b+c+d+1-z)![z!(e+f-a-d+z)!\times$$
$$\times(e+f-b-c+z)!(a+b-e-z)!(c+d-e-z)!\times$$
$$\times(a+c-f-z)!(b+d-f-z)!]^{-1}, \tag{3.15}$$

where $\Delta(a\,b\,c)$ is given by equation (2.34).

In the same way as the general expression for vector addition coefficients (3.15) may be reduced to simple closed forms by assigning a definite value to one argument, say e. Formulae for $e = 0, \frac{1}{2}, 1, 3/2, 2$ have been given by Biedenharn *et al.* [7],

TABLE 4

Formulae for $W(abcd;\,ef)$ with $e = \frac{1}{2}$, and 1

$$W(aa+\tfrac{1}{2}bb+\tfrac{1}{2};\,\tfrac{1}{2}c) = (-)^{a+b-c}\left[\frac{(a+b+c+2)(a+b-c+1)}{(2a+1)(2a+2)(2b+1)(2b+2)}\right]^{\frac{1}{2}}$$

$$W(aa+\tfrac{1}{2}bb-\tfrac{1}{2};\,\tfrac{1}{2}c) = (-)^{a+b-c}\left[\frac{(a-b+c+1)(c-a+b)}{(2a+1)(2a+2)2b(2b+1)}\right]^{\frac{1}{2}}$$

$$W(aa+1bb+1;\,1c) = (-)^{a+b-c}\left[\frac{(a+b+c+3)(a+b+c+2)(a+b-c+2)(a+b-c+1)}{4(2a+3)(a+1)(2a+1)(2b+3)(2b+1)(b+1)}\right]^{\frac{1}{2}}$$

$$W(aa+1bb;\,1c) = (-)^{a+b-c}\left[\frac{(a+b+c+2)(a-b+c+1)(a+b-c+1)(c-a+b)}{4(2a+3)(a+1)(2a+1)b(b+1)(2b+1)}\right]^{\frac{1}{2}}$$

$$W(aa+1bb-1;\,1c) = (-)^{a+b-c}\left[\frac{(c-a+b)(c-a+b-1)(a-b+c+2)(a-b+c+1)}{4(2a+3)(a+1)(2a+1)(2b-1)b(2b+1)}\right]^{\frac{1}{2}}$$

$$W(aabb;\,1c) = (-)^{a+b-c-1}\frac{a(a+1)+b(b+1)-c(c+1)}{[4a(a+1)(2a+1)b(b+1)(2b+1)]^{\frac{1}{2}}}$$

those for $e = 0, \frac{1}{2}, 1$ can be found in Table 4. Extensive numerical tables have been prepared by Biedenharn [4] Simon *et al.* [66], and Rotenberg *et al.* [58].

The unitary nature of the transformation (3.8) which defines the Racah function leads to the orthogonality relation

$$\sum_e(2e+1)(2f+1)\,W(abcd;\,ef)\,W(abcd;\,eg) = \delta_{fg}. \tag{3.16}$$

We can use the expansions (3.6) and (3.7) of the three-vector states to get relationships between the transformation coefficients (3.9). For example, the symmetry properties of the vector addition coefficient (Appendix I) show that

$$|a, (db)\,f;\, c\rangle = (-1)^{f-b-d}\,|a, (bd)\,f;\, c\rangle, \tag{3.17}$$

so that

$$\begin{aligned}\langle a, (db)f;\, c|(ab)e, d;\, c\rangle &= (-1)^{f-b-d}\,\langle a, (bd)f;\, c|(ab)e, d;\, c\rangle \\ &= (-1)^{f-b-d}\,[(2e+1)(2f+1)]^{\frac{1}{2}}\times \\ &\quad\times W(abcd;\, ef)\end{aligned} \tag{3.18}$$

from equation (3.9). Similarly we can show that

$$\langle (bd)f, a;\, c|(ab)e, d;\, c\rangle = (-1)^{c-f-a}\,\langle a, (bd)f;\, c|(ab)e, d;\, c\rangle \tag{3.19}$$

A fruitful source of sum rules is the closure relation for transformation coefficients

$$\langle A|B\rangle = \textstyle\sum_C \langle A|C\rangle\langle C|B\rangle, \tag{3.20}$$

where the sum runs over a complete set of eigenstates $|C\rangle$. Applying this to the change-of-coupling coefficients we can write, for example,

$$\begin{aligned}&\langle a, (bd)\,f;\, c|(ab)\,e, d;\, c\rangle = \\ &\quad= \textstyle\sum_g \langle a, (bd)f;\, c|(ad)g, b;\, c\rangle\langle (ad)g, b;\, c|(ab)e, d;\, c\rangle,\end{aligned}$$

which using equation (3.19) gives the Racah sum rule

$$\textstyle\sum_g (2g+1)(-1)^{p+g}\, W(adcb;\, gf)\, W(abdc;\, eg) = W(abcd;\, ef), \tag{3.21}$$

where $p = a+b+c+d+e+f$. Other sum rules derived in a similar way and some other relations are displayed in Appendix II [5], [7], [23].

Parallel to his treatment of the vector addition coefficients as contraction symbols Wigner [77] has defined a $6-j$ symbol. This differs from the Racah function in phase only,

$$\begin{Bmatrix} a & b & e \\ d & c & f \end{Bmatrix} = (-1)^{a+b+c+d}\, W(abcd;\, ef). \tag{3.22}$$

The $6-j$ symbol has somewhat higher symmetry being invariant under interchange of any two columns and also under the interchange of the upper and lower arguments in each of any two columns (Appendix II). Some extensive tables of the $6-j$ symbol are now available [38], [58].

3.4. Four Angular Momentum Vectors

A state in which four angular momenta a, b, d, and e have a resultant i is specified by giving the resultants of two pairs of the vectors. For example, we may couple a and b to a resultant c, then d and e to f before finally adding c and f to give i. This state with total z-component m can be written as

$$|(ab)c,\ (de)f,\ im\rangle$$

being an eigenfunction of the angular momentum operators

$$\mathbf{A}^2,\ \mathbf{B}^2,\ \mathbf{D}^2,\ \mathbf{E}^2,\ \mathbf{C}^2 = (\mathbf{A}+\mathbf{B})^2,\ \mathbf{F}^2 = (\mathbf{D}+\mathbf{E})^2,$$
$$\mathbf{I}^2 = (\mathbf{A}+\mathbf{B}+\mathbf{D}+\mathbf{E})^2 = (\mathbf{C}+\mathbf{F})^2 \text{ and } I_z.$$

These form a complete set for describing its angular momentum properties.

Clearly there are three ways of choosing pairs from a, b, d, and e, but just as in the three vector case discussed above the corresponding eigenfunctions are not independent. They are connected by a linear transformation, e.g.

$$|(ad)g,\ (be)h;\ im\rangle = \sum_{cf} |(ab)c,\ (de)f;\ im\rangle \times$$
$$\times \langle (ab)c,\ (de)f;\ i|(ad)g,\ (be)h;\ i\rangle. \quad (3.23)$$

The transformation coefficient in equation (3.23) that changes the coupling defines the $9-j$ symbol of Wigner [77].

$$\langle (ab)c,\ (de)f;\ i|(ad)g,\ (be)h;\ i\rangle =$$
$$= [(2c+1)(2f+1)(2g+1)(2h+1)]^{\frac{1}{2}} \begin{Bmatrix} a & b & c \\ d & e & f \\ g & h & i \end{Bmatrix}, \quad (3.24)$$

which is identical to the X-function of Fano [28]. For typographical convenience it is often written $X(abc,\ def,\ ghi)$. We may obtain an expression for the $9-j$ symbol by expanding both sides of equation (3.23) with vector addition coefficients. The calculation is exactly similar to that for the W-function in equations (3.10)–(3.13), thus we shall not give it here. The result expressed as a contraction of six Wigner $3-j$ symbols is given in Appendix III together with some related formulae.

The Racah W-function may be used to contract the sums over vector addition coefficients, leading to the form most suitable for numerical evaluation

$$X(abc,\, def,\, ghi) = \textstyle\sum_k (2k+1)\, W(aidh;\, kg)\times \\ \times W(bfhd;\, ke)\, W(aibf;\, kc). \quad (3.25)$$

There is often a small number of terms only in the sum over k, k being limited by triangular inequalities in the triads (kai), (kbf), and (kdh). In particular, if one of the arguments of X is zero the sum over k reduces to one term (Appendix II).

The orthogonality properties of the $9-j$ symbol follow from the unitary nature of the change of coupling transformation (3.23)

$$\textstyle\sum_{gh} (2c+1)(2f+1)(2g+1)(2h+1) \begin{Bmatrix} a & b & c \\ d & e & f \\ g & h & i \end{Bmatrix} \begin{Bmatrix} a & b & c' \\ d & e & f' \\ g & h & i \end{Bmatrix} = \\ = \delta_{cc'}\delta_{ff'}. \quad (3.26)$$

The $9-j$ symbol is highly symmetrical. Interchange of any two adjacent rows or columns multiplies it by $(-1)^p$ where

$$p = a+b+c+d+e+f+g+h+i,$$

i.e. the sum of all its arguments. It is also invariant under reflection about either diagonal.

The closure relation (3.20) can be used to generate sum rules just as in the case of Racah functions in equation (3.21). In particular, we use

$$\langle (ab)c,\, (de)f; j|(ad)g,\, (be)h; j\rangle = \\ = \textstyle\sum_{kl} \langle (ab)c,\, (de)f; j|(ae)k,\, (bd)l; j\rangle \times \\ \times \langle (ae)k,\, (bd)l; j|(ad)g,\, (be)h; j\rangle,$$

to give the analogue of the Racah sum rule‡ (3.21)

$$\begin{Bmatrix} a & b & c \\ d & e & f \\ g & h & j \end{Bmatrix} = \sum_{kl} (-1)^{h-f-l-2d}(2k+1)(2l+1)\times$$
$$\times \begin{Bmatrix} a & b & c \\ e & d & f \\ k & l & j \end{Bmatrix} \begin{Bmatrix} a & d & g \\ e & b & h \\ k & l & j \end{Bmatrix} \qquad (3.27)$$

Other such relations are left to Appendix III.

The large number of arguments makes tabulation of the $9-j$ difficult, but despite this fairly extensive numerical tables are now available [67]. The $9-j$ symbols with two arguments equal to $\frac{1}{2}$ are of particular importance. They represent the coefficients in the change from $L-S$ to $j-j$ coupling of two spin $-\frac{1}{2}$ particles and have simple explicit forms.

3.5. More Complex Coupling Schemes

The number of possible coupling modes rapidly increases with increase in the number n of angular momentum vectors. Corresponding to transformations between these modes we may define more complex coefficients or invariants. Owing to the difficulty of numerical tabulation however, their usefulness is confined to arguments based on symmetry and orthogonality properties. Two $12-j$ symbols have been defined and their properties discussed [24], [39], [47], [61]; they correspond to changes of coupling of 5 vectors.

For $n \geqslant 4$ it can be shown [Sharp 62] that there exist two $3n-j$ symbols whose symmetries can be displayed by writing their arguments on a cylindrical band, in one case twisted (a Mobius strip), in the other not. When n is greater than 4 there exist other less symmetric symbols as well. Wigner [77] has shown that these coefficients are not specific to the group of real rotations in space with which we are concerned here, but have their analogues for any arbitrary simply reducible group.

‡ We have again used the change of phase that occurs when the order of the vectors is changed, e.g.

$$|(ab)c,\ (ed)f;\ i\rangle = (-1)^{f-e-d}|(ab)c,\ (de)f;\ i\rangle.$$

CHAPTER IV

TENSORS AND TENSOR OPERATORS

4.1. Scalars and Vectors

WE are accustomed to deal with certain physical quantities such as mass and energy which are in no way connected with the orientation of our coordinate system and which have no directional properties. Such quantities are scalars, or tensors of rank zero.‡

Other quantities, such as the position of a point in space or the velocity of a particle, have associated with them a direction as well as a magnitude. These quantities are vectors or tensors of rank one and in order to specify them we must give the direction as well as the magnitude. Alternatively, a vector can be specified by giving its components along the directions *Ox*, *Oy*, *Oz* of a set of orthogonal axes. The components representing the vector depend upon the particular choice of axes, and changing the axes changes the components in a specific way. The components (x_1, x_2, x_3), or (x_i), of the position vector **r** with respect to a new set of axes are obtained from the components (x_1', x_2', x_3'), or (x_i'), referred to the original axes by a linear transformation

$$x_i' = \sum_j a_{ij} x_j. \tag{4.1}$$

The coefficients a_{ij} are the direction cosines of the new axes with respect to the old and are definite functions of the Euler angles specifying the rotation which takes the old axes into the new. If **A** is an arbitrary vector with components (A_i), then these components transform under rotation of axes according to the same law as the components of **r**,

$$A_i' = \sum_j a_{ij} A_j \tag{4.2}$$

‡ Pseudoscalars, which depend upon the handedness of the coordinate system but not its orientation, are counted as scalar for the purpose of this discussion.

where the coefficients a_{ij} are the same as in (4.1). This transformation leads to a new definition of a vector, as a quantity represented by three components which transform according to (4.2) when the coordinate axes are changed.

Let us consider a property of a physical system represented by a vector **A** with components (A_i) with respect to some chosen set of axes, and enquire how the vector **A** transforms when the system is rotated. The discussion of section 1.4 shows that the change in **A** produced by rotating the system can be described by making an opposite rotation of the coordinate axes. Thus **A** transforms to a vector **A**′ with components

$$A_i' = \sum_j \bar{a}_{ij} A_j.$$

The transformation matrix $\bar{a}_{ij}$ is the inverse of the matrix a_{ij} of (4.2) for the same rotation.

On either view of a rotation, considering it as a rotation of a physical system or as a coordinate transformation, the new components of a vector are given in terms of the old by a matrix transformation which is specified by the Euler angles of the rotation. These matrices form a representation of the rotation group of dimension 3, and the components of the vector are the basis for the representation.

4.2. Tensors of Higher Rank

The mass moments of a system of particles are given by

$$M_{ij} = \sum_\alpha m_\alpha x_{\alpha i} x_{\alpha j},$$

where $(x_{\alpha i})$ is the position vector of the particle α and m_α is its mass. The mass moments are a set of 6 linearly independent quantities $(M_{ij} = M_{ji})$ and represent a symmetric tensor of rank 2. On rotation of axes the particle coordinates transform as in equation (4.1) hence the mass moments transform as

$$M_{ij}' = \sum_{kl} a_{ik} a_{jl} M_{kl}. \tag{4.3}$$

A general (non-symmetric) tensor of rank 2 is represented by

9 components T_{ij} with the same transformation law on rotation of axes as equation (4.3)

$$T'_{ij} = \sum_{kl} a_{ik} a_{jl} T_{kl}. \tag{4.4}$$

Similarly the general tensor of rank n is represented by 3^n components which transform according to the obvious generalization of equation (4.4).

If the nine components T_{ij} of a general second-rank-tensor are written as a column vector, the transformation coefficients form a 9×9 matrix; the set of these matrices for all rotations forms a representation of the rotation group of dimension 9. It is well-known that one can form from the general second-rank tensor T_{ij} a scalar $\sum_k T_{kk}$, an antisymmetric tensor

$$\hat{T}_{ij} = \tfrac{1}{2}(T_{ij} - T_{ji}),$$

and a symmetric tensor with zero trace,

$$\bar{T}_{ij} = \tfrac{1}{2}(T_{ij} + T_{ji}) - \tfrac{1}{3}\delta_{ij} \sum_k T_{kk}.$$

Conversely the 5 independent components $\bar{T}_{ij}$, the 3 independent components $\hat{T}_{ij}$, and $\sum_k T_{kk}$ together specify the nine independent components T_{ij}. Under a rotation the components $\hat{T}_{ij}$ transform amongst themselves as do the components $\bar{T}_{ij}$, while $\sum_k T_{kk}$ is invariant. Thus the above reduction of the general second-rank tensor T_{ij} corresponds to the reduction of the 9 dimensional representation of the rotation group of equation (4.4) to its irreducible components of dimension 5, 3, and 1. The tensors $\bar{T}_{ij}$, $\hat{T}_{ij}$ and $\sum_k T_{kk}$ are the irreducible components of the general second rank tensor belonging to the representations $\mathscr{D}_2$, $\mathscr{D}_1$, and $\mathscr{D}_0$ of the rotation group.

4.3. Irreducible Spherical Tensors

Cartesian tensors are defined in section 4.2 as quantities represented by a set of components which have a definite transformation law under rotations of the coordinate system. It is natural to define a general spherical tensor $\mathbf{T}_k$ of rank k as a quantity represented by $2k+1$ components T_{kq} which transform according to the irreducible representation $\mathscr{D}_k$ of the rotation group.‡

$$T'_{kq} = \sum_p T_{kp}\, \mathscr{D}^k_{pq}(\alpha\beta\gamma) \tag{4.5}$$

$(\alpha\beta\gamma)$ are the Euler angles (section 2.4) of the rotation taking the old, unprimed, axes into the new, primed, axes. Thus (4.5) expresses a component T'_{kq} with respect to the new axes in terms of the components T_{kq} defined with respect to the old axes.

It is an immediate consequence of the irreducibility of $\mathscr{D}_k$ that the tensor $\mathbf{T}_k$ is irreducible.

4.4. Products of Tensors

In the theory of Cartesian tensors the simple (uncontracted) product of two tensors of rank m and n respectively yields a tensor of tank $m+n$. For example, the nine products $a_i b_j$ of the components of two vector **a** and **b** are the components of a second rank tensor, which transforms according to the representation $\mathscr{D}_1 \times \mathscr{D}_1$ of the rotation group. The tensor $a_i b_j$ gives rise to irreducible tensors $\frac{1}{2}(a_i b_j + a_j b_i) - \frac{1}{3}\delta_{ij}\mathbf{a}\,.\,\mathbf{b}$, $\mathbf{a} \wedge \mathbf{b}$ and $\mathbf{a}\,.\,\mathbf{b}$, and this reduction corresponds to the reduction

$$\mathscr{D}_1 \times \mathscr{D}_1 = \mathscr{D}_2 + \mathscr{D}_1 + \mathscr{D}_0$$

of the rotation group (equation (2.30)).

If **c** is a third different vector, the set of 27 products $a_i b_j c_k$ represent a third rank tensor transforming according to the representation $\mathscr{D}_1 \times \mathscr{D}_1 \times \mathscr{D}_1$ of the rotation group. This representation is reducible as follows,

$$\begin{aligned} \mathscr{D}_1 \times (\mathscr{D}_1 \times \mathscr{D}_1) &= \mathscr{D}_1 \times (\mathscr{D}_2 + \mathscr{D}_1 + \mathscr{D}_0), \\ &= \mathscr{D}_3 + 2\mathscr{D}_2 + 3\mathscr{D}_1 + \mathscr{D}_0; \end{aligned}$$

that is, the 27 components of the tensor $a_i b_j c_k$ give rise to one irreducible tensor of rank 3, two of rank 2, three vectors and

‡ The definition (4.5) is chosen to agree with the transformation law (2.24) for spherical harmonics on rotation of axes. This means that spherical harmonics provide a special example of spherical tensors.

one scalar. Explicitly the three vectors are $(\mathbf{a}.\mathbf{b})\mathbf{c}$, $(\mathbf{b}.\mathbf{c})\mathbf{a}$, $(\mathbf{c}.\mathbf{a})\mathbf{b}$, and the scalar is $(\mathbf{a}\wedge\mathbf{b}).\mathbf{c}$. In this way we can count the number of different irreducible parts of a general cartesian tensor of rank n.

If R_{kq} and $S_{k'q'}$ are irreducible tensors of rank k and k' respectively in the spherical representation, the $(2k+1)(2k'+1)$ products $R_{kq}S_{k'q'}$ form a tensor transforming under the representation $\mathscr{D}_k \times \mathscr{D}_{k'}$, of the rotation group. This representation is reducible (equation (2.30)) and its reduction gives the irreducible tensors‡

$$T_{KQ}(kk') = \sum_{qq'} R_{kq}S_{k'q'}\langle kk'qq'|KQ\rangle, \tag{4.6}$$

with K running from $k+k'$ to $|k-k'|$ and $Q = q+q'$.

Unless the operators R_k and S'_k commute there is no simple relation between $T_{KQ}(R_k, S_{k'})$ and $T_{KQ}(S_{k'}, R_k)$. An interesting special case occurs when $R = S$ and $k = k'$ which is analogous to the exchange symmetry property of two-particle states (section 2.7.3). Provided the components R_{kq} and R_{kQ-q} commute interchanging them merely multiplies $T_{KQ}(kk)$ by $(-1)^{K-2k}$ $(=(-1)^K$ if k is integral). Thus only product tensors with K even do not vanish. This property is a generalization of the vector relation $\mathbf{v}\wedge\mathbf{v} = 0$.§

When $k = k'$ and $K = 0$ the product (4.6) is the generalization of the scalar product of two vectors. When k is integral another phase and normalization is used in the definition of this product

$$\mathbf{R}_k.\mathbf{S}_k = \sum_q (-1)^q R_{kq}S_{k-q} = (-1)^k\sqrt{(2k+1)}\; T_{00}(\mathbf{R},\mathbf{S}). \tag{4.7}$$

4.5. Tensor Operators

The notion of tensors can be extended directly to quantum mechanical tensor operators. An irreducible tensor operator

‡ Sometimes we use as arguments of a composite tensor the actual tensors out of which it is constructed writing it as

$$T_{KQ}(R_k, S_{k'}) \quad \text{instead of} \quad T_{KQ}(k, k').$$

§ The angular momentum J does not obey this relation but rather $\mathbf{J}\wedge\mathbf{J} = i\mathbf{J}$, because its components do not commute.

of rank k is an operator with $2k+1$ components T_{kq} which transform under a rotation $(\alpha\beta\gamma)$ of axes as‡

$$T'_{kq} = DT_{kq}\,D^{+} = \sum_{p} T_{kp}\mathscr{D}^{k}_{pq}(\alpha\beta\gamma). \tag{4.8}$$

Relation (4.8) with D representing an infinitesimal rotation leads to commutation laws of the components of the angular momentum $\mathbf{J}$ with the components of $\mathbf{T}_k$.

Let D be the infinitesimal rotation $(1-i\alpha J_\lambda)$. Thus from equation (2.15)

$$\mathscr{D}^{k}_{pq} = \langle kp|1-i\alpha J_\lambda|kq\rangle = \delta_{pq}-i\alpha\langle kp|J_\lambda|kq\rangle.$$

With this rotation equation (4.8) becomes

$$(1-i\alpha J_\lambda)T_{kq}(1+i\alpha J_\lambda) = \sum_{p} T_{kp}\mathscr{D}^{k}_{pq},$$

or

$$J_\lambda T_{kq}-T_{kq}\,J_\lambda = \sum_{p} T_{kp}\langle kp|J_\lambda|kq\rangle.$$

Putting $J_\lambda = J_z$ and $J_{\pm}$ in turn and using the matrix elements of J_λ from equations (2.4) one obtains

$$\begin{aligned}[J_z,\, T_{kq}] &= qT_{kq},\\ [J_{\pm},\, T_{kq}] &= [(k\pm q+1)(k\mp q)]^{\frac{1}{2}}T_{kq\pm 1}.\end{aligned} \tag{4.9}$$

The commutation rules (4.9) of $\mathbf{J}$ with spherical tensor operators can be used in finding the spherical equivalents of Cartesian tensors. For example, if $\mathbf{A}$ is a vector,

$$[J_z,\, A_z] = 0$$

from the commutation relations in equations (1.9). Thus $A_0 = A_z$. From (4.9) we have

$$\begin{aligned}A_{\pm 1} &= \frac{1}{\sqrt{2}}[J_{\pm},\, A_0],\\ &= \frac{1}{\sqrt{2}}[J_x,\, A_z]\pm i[J_y,\, A_z],\\ &= \mp\frac{1}{\sqrt{2}}(A_x\pm iA_y).\end{aligned} \tag{4.10}$$

‡ A rotation R of a quantum system transforms an operator A as

$$A' = D^{+}(R)\,AD\,(R), \text{ cf. section 1.5.}$$

However, a rotation R of axes is equivalent to an inverse rotation R^{-1} of the

Again, if **A** and **B** are vectors the spherical tensor of rank 2 formed by their product is

$$T(AB)_{2q} = \sum_{mn} A_m B_n \langle 11mn|2q \rangle.$$

In particular, using the value of $\langle 22|1111 \rangle = 1$

$$T(AB)_{22} = A_1 B_1.$$

Then from the commutation relations (4.9)

$$\begin{aligned} T(AB)_{21} &= \tfrac{1}{2}[J_-, A_1 B_1] \\ &= \tfrac{1}{2}([J_-, A_1]B_1 + A_1[J_-, B_1]) \\ &= \frac{1}{\sqrt{2}}(A_0 B_1 + A_1 B_0) \end{aligned} \tag{4.11}$$

and $$T(AB)_{20} = \frac{1}{\sqrt{6}}(3A_z B_z - \mathbf{A} \cdot \mathbf{B}).$$

4.6. Spherical Harmonics as Tensors

Spherical harmonics transform on rotation of axes according to equation (4.5), hence are examples of spherical tensors. There are some relations involving products of spherical harmonics which follow simply from section 4.4. These take their simplest form when expressed in terms of the modified spherical harmonic

$$C_{kq} = \left(\frac{4\pi}{2k+1}\right)^{\frac{1}{2}} Y_{kq}.$$

If $C_{kq}(\theta, \varphi)$ and $C_{kq}(\theta', \phi')$ are modified spherical harmonics then the scalar product (equation (4.7))

$$\mathbf{C}_k \cdot \mathbf{C}'_k = \sum_q (-1)^q C_{kq}(\theta, \varphi) C_{k-q}(\theta', \varphi'),$$

is invariant with respect to rotation of axes. It follows that the scalar product must be a function of the angle Θ between the directions (θ, φ) and (θ', ϕ'), this angle being the only quantity

system so that the transformation operator in equation (4.8) is $D(R^{-1})$ or $D^+(R)$ rather than $D(R)$. If $|0\rangle$ is a spherically symmetric wave function $T_{kq}|0\rangle$ is a wave function with angular momentum quantum numbers (k, q).

independent of choice of axes. Choosing axes so that the direction (θ', ϕ') becomes the new z-axis

$$C_{kq}(\theta', \phi') \to C_{kq}(0, 0) = \delta(q, 0), \text{ (cf. equation (2.10))}$$

$$C_{k0}(\theta, \phi) \to P_k(\cos\theta) = P_k(\cos\Theta), \quad \text{(cf. equation (2.11))}$$

and the scalar product (cf. equation (2.25))

$$\mathbf{C}_k \cdot \mathbf{C}'_k = P_k(\cos\Theta). \tag{4.12}$$

Again, if $C_{kq}(\theta,\varphi)$ and $C_{k'q'}(\theta, \varphi)$ are spherical harmonics of the same angles (θ, φ) then

$$\sum_{qq'} \langle KQ|kk'qq'\rangle C_{kq} C_{k'q'},$$

is a tensor of rank K. Since it is a function of (θ, φ) only it must be proportional to $C_{KQ}(\theta, \phi)$. That is

$$\sum_{qq'} \langle KQ|kk'qq'\rangle\, C_{kq}(\theta, \phi) C_{k'q'}(\theta, \phi) = A_K C_{KQ}(\theta, \varphi),$$

$$= \langle KO|kk'00\rangle\, C_{KQ}(\theta, \phi). \tag{4.13}$$

The value of A_K in equation (4.13) is found by putting $\theta = \phi = 0$ and making use of equation (2.10). Equation (4.13) follows also as a special case of the rule (2.31) for the combination of rotation matrices.

Examples of the spherical harmonics polarized by the vector operators $\mathbf{\nabla}$ and $\mathbf{L}$ are discussed in section 4.10.2 in relation to vector fields.

Spherical harmonics with different arguments may also be used to construct product tensors

$$B_{KQ}(kk') = \sum_{qq'} \langle KQ|kk'qq'\rangle\, C_{kq}(\mathbf{u}) C_{k'q'}(\mathbf{u}'). \tag{4.14}$$

These tensors, which we may call bipolar harmonics, continually appear in problems involving two directions, ($\mathbf{u}$ and $\mathbf{u}'$ being unit vectors along those directions). For example, they describe the angle dependence of two particles moving in a central field in an eigenstate of total angular momentum L,

$$\langle \mathbf{u}, \mathbf{u}'|ll'LM\rangle = [(2l+1)(2l'+1)]^{\frac{1}{2}} B_{LM}(ll')/4\pi.$$

Often they are required in problems of the angular correlation of nuclear radiations [5], [16].

The bipolar harmonics are orthogonal for integration over the angles of $\mathbf{u}$ and $\mathbf{u}'$

$$\int d\mathbf{u}\int d\mathbf{u}'\, B^*_{LM}(l_1 l_2) B_{L'M'}(l'_1 l'_2) = 16\pi^2 \frac{\delta(l_1 l'_1)\,\delta(l_2 l'_2)\,\delta(LL')\,\delta(MM')}{(2l_1+1)(2l_2+1)}$$

(they would also be normalized to unity if they had been defined with the normalized Y_{kq} rather than with the C_{kq}) and have the closure property

$$\sum_{LM} |B_{LM}(l_1 l_2)|^2 = 1.$$

When $K = Q = 0$, (4.14) reduces to the spherical harmonic addition theorem (4.12)

$$B_{00}(kk') = (-)^k\,\delta(kk')P_k(\mathbf{u}\,.\,\mathbf{u}')/\sqrt{(2k+1)}.$$

Again, when $k = k' = 1$, we retrieve the vector relations

$$B_{00}(11) = -\mathbf{u}\,.\,\mathbf{u}'/\sqrt{3}$$

and

$$B_{1q}(11) = i(\mathbf{u}\wedge\mathbf{u}')_q/\sqrt{2}.$$

4.7. Matrix Elements of Tensor Operators

4.7.1. Wigner–Eckart Theorem

We wish to evaluate matrix elements of a tensor operator with respect to the state vectors of a dynamical system. When angular momentum is conserved, so that the state vectors are eigenfunctions of $\mathbf{J}^2$ and $\mathbf{J}_z$ the matrix elements of a tensor operator have a simple geometrical dependence on the magnetic quantum numbers. Let T_{kq} be a tensor operator of rank k and consider the matrix element $\langle\alpha JM|T_{kq}|\alpha' J'M'\rangle$. The state vector $T_{kq}|\alpha' J'M'\rangle$ transforms according to the representation $\mathscr{D}_k\times\mathscr{D}_{J'}$ of the rotation group. We reduce this representation to its irreducible components by forming the vectors with angular momentum (K, Q)

$$|\beta KQ\rangle = \sum_{qM'} \langle KQ|J'kM'q\rangle T_{kq}|\alpha' J'M'\rangle.$$

Inverting this relation and taking matrix elements with $\langle \alpha JM|$ gives

$$\begin{aligned}\langle \alpha JM|T_{kq}|\alpha' J' M'\rangle &= \sum_{KQ} \langle \alpha JM|\beta KQ\rangle\langle KQ|J'kM'q\rangle \\ &= \langle \alpha JM|\beta JM\rangle\langle JM|J'kM'q\rangle \\ &= (-1)^{2k}\langle \alpha J||\mathbf{T}_k||\alpha' J'\rangle\langle JM|J'kM'q\rangle. \quad (4.15)\end{aligned}$$

The inner product of state vectors $\langle \alpha JM|\beta JM\rangle$ is independent of M. Thus the matrix element of the tensor operator factorizes into two parts. The directional properties are contained in the Clebsch-Gordan coefficients and the dynamics of the system appear only in the scalar matrix element $\langle \alpha JM|\beta JM\rangle$ usually written as $\langle \alpha J||\mathbf{T}_k||\alpha' J'\rangle$ and called the reduced matrix element.‡

Inverting (4.15) gives the analogue to the scalar contraction of vectors (section 3.2)

$$\langle \alpha J||\mathbf{T}_k||\alpha' J'\rangle = (-1)^{2k} \sum_{M'q} \langle JM|J'kM'q\rangle\langle \alpha JM|T_{kq}|\alpha' J' M'\rangle.$$

The prototype of such matrix elements is the integration over three spherical harmonics:

$$\langle lm|\,Y_{LM}|l'm'\rangle = \int Y_{lm}(\theta,\phi)^*\; Y_{LM}(\theta,\varphi)\; Y_{l'm'}(\theta,\varphi)d\Omega. \quad (4.16)$$

This is evaluated by combining the last two harmonics in the way described in the last section (equation (4.13))

$$Y_{LM} Y_{l'm'} = \sum_{kq} Y_{kq}(\theta,\varphi)\langle kq|Ll'Mm'\rangle\langle k0|Ll'00\rangle \times \times\left[\frac{(2l'+1)(2L+1)}{4\pi(2k+1)}\right]^{\frac{1}{2}}.$$

From the orthogonality properties of the harmonics we see that only the $k = l$, $m = q$ term has a non-zero integral. Thus

$$\langle lm|\,Y_{LM}|l'm'\rangle = \langle lm|l'Lm'M\rangle\langle l||\,Y_L||l'\rangle$$

where

$$\langle l||\,Y_L||l'\rangle = \left[\frac{(2l'+1)(2L+1)}{4\pi(2l+1)}\right]^{\frac{1}{2}}\langle l0|Ll'00\rangle. \quad (4.17)$$

‡ The definition used here is that of Wigner [78] and Rose [54]. Wigner's tensors $T^k_{(q)}$, however, are equivalent to the adjoints, T^+_{kq} of ours (cf. section 4.8) because of the rotation conventions he adopts (Appendix IV). Racah [48], [31], and Edmonds [22] define reduced elements which are $(2J+1)^{\frac{1}{2}}$ times those in (4.15). (The factor $(-)^{2k}$ makes the phases of the two definitions identical.)

The physical significance of the Wigner-Eckart theorem is now clear. The 'integrand' of the matrix element on the left of (4.16) must be a scalar if the element is not to vanish. So the only part of the product of T_{kq} and $|J'M'\rangle$ which can contribute is that which rotates contravariantly to $\langle JM|$; that is, like $|JM\rangle$. By definition its amplitude is just the Clebsch–Gordan coefficient, and this is the geometrical factor containing all dependence on magnetic quantum numbers. A further important example is the matrix of the angular momentum operator **J** whose spherical components J_λ form a tensor of rank 1 as discussed in section 4.5. The reduced matrix is easily obtained by considering $J_0 = J_z$:

$$\begin{aligned}\langle JM|J_z|J'M'\rangle &= M\delta(JJ')\,\delta(MM')\\ &= \langle JM|J'1M'0\rangle\langle J||\mathbf{J}||J'\rangle.\end{aligned}$$

The explicit form for the Clebsch–Gordan coefficient (Table 3) gives immediately

$$\langle J'||\mathbf{J}||J\rangle = \delta(JJ')\sqrt{\{J(J+1)\}}. \tag{4.18}$$

All processes which do not involve a definite spatial direction (such as radiation transition probabilities) are independent of the magnetic quantum numbers and are described by reduced matrix elements. This can be seen explicitly by supposing that matrix element $\langle JM|T_{kq}|J'M'\rangle$ describes a radiative transition from state $|JM\rangle$ to $|J'M'\rangle$. If there is no preferred direction then all orientations M' of the final state are equivalent and the total transition probability is

$$\begin{aligned}&\sum_{qM'}|\langle JM|T_{kq}|J'M'\rangle|^2\\ &\quad= |\langle J||T_k||J'\rangle|^2\sum_{qM'}|\langle JM|kJ'qM'\rangle|^2\\ &\quad= |\langle J||T_k||J'\rangle|^2\end{aligned}$$

since the Clebsch–Gordan coefficients are normalized.

The above theorem due to Wigner and Eckart [78], [21] represents an important step in the program outlined in the introduction to make a division between the geometrical and physical properties of a system, or to remove from the problem those aspects which relate to the symmetry of the system. The Wigner–Eckart theorem makes this division explicit, the matrix elements factorizing into a Clebsch–Gordan coefficient containing purely geometrical information concerning the orientation of the system and a reduced matrix element depending on the detailed physical structure of the system.

4.7.2. Selection Rules

Selection rules connected with the conservation of angular momentum are mostly derived by using the Wigner–Eckart theorem. It follows from the properties of the Clebsch–Gordan coefficient in section 2.7 that the matrix element

$$\langle JM|T_{kq}|J'M'\rangle$$

vanishes unless

$$M = q+M'$$

and the triangular conditions are satisfied,

$$|J-J'| \leqslant k \leqslant J+J'. \tag{4.19}$$

Clearly for $k = 0$ we must have $J = J'$, $M = M'$ and the corresponding Clebsch–Gordan coefficient is unity. This shows that the matrix element of a scalar is independent of the magnetic quantum number (i.e. of the choice of quantization axis), as it must be, and is identical to the reduced matrix element with our definition

$$\langle JM|T_{00}|J'M'\rangle = \delta(JJ')\ \delta(MM')\ \langle J||T_0||J'\rangle.$$

The spherical harmonic integral (4.16) also contains the parity selection rule for the spatial part of a single particle matrix element; $(l+l'+L)$ must be even, otherwise $\langle ll'00|L0\rangle$ vanishes (cf. equation (2.35)).

4.8. The Adjoint of a Tensor Operator

The hermitian conjugate or adjoint T^+ of an operator T is defined by expressing its matrix elements in terms of matrix elements of T as

$$\langle 1|T^+|2\rangle = \langle 2|T|1\rangle^*. \tag{4.20}$$

The operator T is called hermitian if $T = T^+$. A necessary and sufficient condition for an operator to be hermitian is that all its eigenvalues should be real. For example, the components of J_x, J_y, J_z of the angular momentum operator are hermitian and from section 2.2 they have real eigenvalues.

If A and B are operators and a is a complex number it follows from the definition (4.20) of the adjoint operator that

$$(AB)^+ = B^+A^+ \quad \text{and} \quad (aA)^+ = a^*A^+ \tag{4.21}$$

hence

$$[A, B]^+ = [B^+, A^+] = -[A^+, B^+].$$

From (4.21) and from the hermitian property of the components of $\mathbf{J}$ it follows that

$$J_\pm^+ = J_\mp, \; J_0^+ = J_0,$$

and from the commutation relations (4.9) of the shift operators $J_\pm$ and J_0 with the components of a spherical tensor operator $\mathbf{T}_k$ that

$$\begin{aligned} [J_0, T_{kq}^+] &= -[J_0, T_{kq}]^+, = -qT_{kq}^+ \\ [J_\pm, T_{kq}^+] &= -[J_\mp, T_{kq}]^+ = -[(k\mp q+1)(k\pm q)]^{\frac{1}{2}} T_{kq\mp 1}^+. \end{aligned}$$

Thus $\mathbf{T}_k^+$ transforms contragradiently to $\mathbf{T}_k$ (section 2.6). Hence if we define an operator $\bar{T}_{kq}$ by

$$\bar{T}_{kq} = (-1)^{p-q} T_{k-q}^+ \tag{4.22}$$

then the $2k+1$ components of $\bar{T}_{kq}$ transform as a tensor operator of rank k, (p is an arbitrary integer or half integer depending on whether k is integral or half integral).

As a result of the property (4.22) at most only the $q = 0$ component of a tensor operator can be hermitian. It is possible, however, to extend the notion of hermitian operators to cover tensor operators by defining a hermitian tensor operator as one with the property

$$T_{kq} = (-1)^{p-q} T_{k-q}^+. \tag{4.23}$$

The choice of phase p is arbitrary, although some authors demand that $p = k$ (Edmonds [22]) and others use $p = 0$ (Schwinger [60]). Then operators which have $p = k+1$

or 1 respectively are called anti-hermitian. On the other hand the phase $(-1)^q$ is essential to preserve the correct rotational properties (4.8). The spherical harmonics Y_{LM} and the spherical components J_q of $\mathbf{J}$ ($J_0 = J_z, J_{\pm 1} = \mp\frac{1}{\sqrt{2}}(J_x \pm iJ_y)$, cf. equation (4.10)) show this hermitian property with $p = 0$.

$$Y^+_{LM} = Y^*_{LM} = (-1)^M Y_{L-M}$$
$$J^+_q = (-1)^q J_{-q}.$$

There is a simple conjugation property for reduced matrix elements of a hermitian tensor operator which corresponds to equation (4.20). From equations (4.22) and (4.20)

$$\begin{aligned}\langle JM|T_{kq}|J'M'\rangle &= (-1)^{p-q}\langle JM|T^+_{k-q}|J'M'\rangle\\ &= (-1)^{p-q}\langle J'M'|T_{k-q}|JM\rangle^*,\end{aligned}$$

or in terms of reduced matrix elements

$$\begin{aligned}&\sqrt{(2J+1)}\,\langle J||T_k||J'\rangle\\ &\qquad = (-1)^{J-J'-p}\sqrt{(2J'+1)}\,\langle J'||T_k||J\rangle^*. \qquad (4.24)\end{aligned}$$

4.9. Time Reversal

The results of section 1.8 do not apply directly to a system with angular momentum because $\mathbf{J}$ is not invariant but changes sign on time reversal. If $\boldsymbol{\theta}$ is the time reversal operator then

$$\boldsymbol{\theta} J_z \boldsymbol{\theta}^{-1} = -J_z$$
$$\boldsymbol{\theta} J_\pm \boldsymbol{\theta}^{-1} = -J_\mp.$$

(The anti-linear character of $\boldsymbol{\theta}$ requires that the complex conjugate of any number which appears in the transformed expressions should be taken, hence $J_+ \to -J_-$.) It follows from the above transformation equations that if $|jm\rangle$ is a set of states with angular momentum (j, m) then

$$\begin{aligned}J_0\,\boldsymbol{\theta}|\alpha jm\rangle &= -\boldsymbol{\theta} J_0|\alpha jm\rangle = -m\boldsymbol{\theta}|\alpha jm\rangle\\ J_\pm\,\boldsymbol{\theta}|\alpha jm\rangle &= -\boldsymbol{\theta} J_\mp|\alpha jm\rangle\\ &= -[(j\mp m+1)(j\pm m)]^{\frac{1}{2}}\,\boldsymbol{\theta}|\alpha jm\mp 1\rangle. \qquad (4.25)\end{aligned}$$

Thus $\boldsymbol{\theta}|\alpha jm\rangle$ transforms contragradiently to $|\alpha jm\rangle$. If we put $|\beta jm\rangle = (-1)^{p-m}\boldsymbol{\theta}|\alpha j-m\rangle$ then $|\beta jm\rangle$ has the correct transformation laws for a wave function of angular momentum (j, m).

We consider a system with Hamiltonian H which is time reversal invariant. The requirement that wave-functions are eigen-states of angular momentum makes it impossible to choose them invariant under time reversal, but the transformation properties of a time-reversed wave function discussed in the last paragraph make it possible to choose a set of eigenstates of H, $\mathbf{J}^2$ and J_z for which

$$\boldsymbol{\theta}|\alpha jm\rangle = (-1)^{p-m}|\alpha j-m\rangle. \tag{4.26}$$

The phase p is so far not specified but it is usual to require that $p = j$, for then the choice of phases of the vector addition coefficients (section 2.3) determines that

$$|JM\rangle = \sum_{m_1 m_2} |j_1 j_2 m_1 m_2\rangle\langle m_1 m_2|JM\rangle$$

satisfies equation (4.26) if it is satisfied by $|j_1 m_1\rangle$ and $j_2 m_2\rangle$. If we say that a tensor operator $\mathbf{T}_k$ has a definite transformation under time reversal when

$$\boldsymbol{\theta}T_{kq}\boldsymbol{\theta}^{-1} = (-1)^{p-q}T_{k-q} \tag{4.27}$$

then by an argument similar to that given in equations (1.13) and (1.14) it can be shown that the reduced matrix elements of $\mathbf{T}_k$ are real if $(p-k)$ is even and pure imaginary if $(p-k)$ is odd. In each case only one real parameter is necessary to specify the value of the matrix element.

The requirement that (4.26) and (4.27) should be satisfied with $p = l$ is the origin of the factor i^l sometimes introduced into the definition of spherical harmonics Y_{lm} (section 2.3).

Equation (4.26) implies that

$$\begin{aligned}\boldsymbol{\theta}^2|\alpha jm\rangle &= (-1)^{2m}|\alpha jm\rangle = (-1)^{2j}|\alpha jm\rangle \\ &= |\alpha jm\rangle \qquad \text{if } j \text{ is an integer} \\ &= -|\alpha jm\rangle \qquad \text{if } j \text{ is a half-integer.}\end{aligned}$$

4.10. Multipole Expansions

4.10.1. Scalar Fields

Let $V(r, \theta, \varphi)$ be an arbitrary scalar field. Rotating the field through a small angle α about the z-axis produces a new field V' given by

$$\begin{aligned} V'(r, \theta, \varphi) &= V(r, \theta, \varphi - \alpha) \\ &= \left(1 - \alpha\frac{\partial}{\partial\phi}\right) V \\ &= (1 - i\alpha L_z)V. \end{aligned} \tag{4.28}$$

The operator $L_z = -i\dfrac{\partial}{\partial\phi} = -i\left(x\dfrac{\partial}{\partial y} - y\dfrac{\partial}{\partial x}\right)$ is the infinitesimal rotation operator for the field about the z-axis. The remaining components of $\mathbf{L}$, L_x and L_y are defined similarly.

A multipole expansion aims at expressing the field V as a sum of components with rotational properties of spherical tensors, and for a scalar field this is achieved by expanding V as a series of spherical harmonics:

$$V(r, \theta, \varphi) = \sum_m V_{lm}(r) Y_{lm}(\theta, \phi)^* \tag{4.29}$$

where $$V_{lm} = \int Y_{lm}(\theta, \phi) V(r, \theta, \phi)\, d\Omega.$$

A rotation of the field $V(r, \theta, \phi)$ specified by Euler angles (α, β, γ) is given by transforming Y_{lm} according to equation (2.22) while a transformation of the field produced by a rotation of axes through angles (α, β, γ) is given by transforming the coefficients V_{lm} according to equation (4.5). A simultaneous transformation of Y_{lm} and V_{lm} corresponds to a rotation of axes with the field and produces no change in V. The components of V_{lm} transform as components of a tensor of rank l and are the multipole components of the field V.

As examples we consider the expansion of a plane wave and of a potential function. A plane wave travelling in the z-direction is symmetrical about the z-axis and can be expanded as a series of Legendre polynomials referred to this axis,

$$e^{ikz} = \sum_l i^l(2l+1) j_l(kr) P_l(\cos\theta). \tag{4.30}$$

The radial functions $j_l(kr)$ can be expressed as Bessel functions of half-odd integer order

$$j_l(kr) = \left(\frac{\pi}{2kr}\right)^{\frac{1}{2}} J_{l+\frac{1}{2}}(kr).$$

We may obtain the equation for a plane wave travelling in an arbitrary direction (β, α) by rotating the wave travelling in the z-direction. Equations (4.30) and (4.5) (or alternatively the addition theorem (4.12) for spherical harmonics) give

$$\exp i\mathbf{k}\,.\,\mathbf{r} = \sum_{lm} i^l(2l+1)j_l(kr)C_{lm}(\theta, \phi)C^*_{lm}(\beta, \alpha). \tag{4.31}$$

The electrostatic field produced by a system of charges provides an example of a scalar potential field. If $\mathbf{r}$ and $\mathbf{r}'$ are two vectors with direction (θ, φ) and (θ', ϕ') and if Θ is the angle between the two vectors then for $r > r'$

$$\frac{1}{|\mathbf{r}-\mathbf{r}'|} = \sum_l \frac{r'^l}{r^{l+1}} P_l(\cos\Theta)$$

$$= \sum_{lm} \frac{r'^l}{r^{l+1}} C_{lm}(\theta, \varphi)C^*_{lm}(\theta', \phi'). \tag{4.32}$$

Thus the potential at $\mathbf{r}$ due to a charge distribution of density $\rho(\mathbf{r}')$ is given by

$$V(\mathbf{r}) = \int \frac{\rho(\mathbf{r}')}{|r-r'|}\,d\mathbf{r}' = \sum_{lm} \frac{Q^*_{lm}}{r^{l+1}} C_{lm}(\theta, \phi), \tag{4.33}$$

where

$$Q_{lm} = \int \rho(\mathbf{r}')r'^l C_{lm}(\theta', \phi')\,d\mathbf{r}'.$$

The tensors Q_{lm} are the multipole moments of the charge distribution. For a system of point charges

$$Q_{lm} = \sum_i e_i r_i^l C_{lm}(\theta_i, \phi_i), \tag{4.34}$$

where (r_i, θ_i, ϕ_i) are the spherical coordinates of the ith particle and e_i is its charge. For a quantum system the lth multipole moment of a state is the expectation value of Q_{lm} in that state.

4.10.2. Vector Fields

Let $\mathbf{A}(r, \theta, \phi) = \sum_i A_i \mathbf{e}_i$ be a vector field. The $\mathbf{e}_i$ are unit vectors along the coordinate axes and the A_i are the components of $\mathbf{A}$ along those axes. A rotation of the field through a small angle α about the z-axis produces a new field

$$\mathbf{A}'(r, \theta, \varphi) = \sum_i A_i(r, \theta, \phi - \alpha)\mathbf{e}'_i$$

where $\mathbf{e}'_i$ are unit vectors along a set of axes rotated with the field. These unit vectors expressed in terms of $\mathbf{e}_i$ are

$$\mathbf{e}'_x = \mathbf{e}_x + \alpha\mathbf{e}_y,\ \mathbf{e}'_y = \mathbf{e}_y - \alpha\mathbf{e}_x \quad \text{and} \quad \mathbf{e}'_z = \mathbf{e}_z$$

or
$$\mathbf{e}'_i = \mathbf{e}_i + \alpha\mathbf{e}_z \wedge \mathbf{e}_i.$$

Expanding the A_i as a power series in α and substituting for $\mathbf{e}'_i$ in terms of $\mathbf{e}_i$ we obtain an expression for the rotated field to the first order in α

$$\mathbf{A}'(r, \theta, \phi) = \mathbf{A} + \alpha\left(\mathbf{e}_z \wedge \mathbf{A} - \frac{\partial \mathbf{A}}{\partial \phi}\right),$$

or
$$\mathbf{A}'(r, \theta, \phi) = (1 - i\alpha J_z)\mathbf{A}, \tag{4.35}$$

where
$$\begin{aligned} J_z &= -i\frac{\partial}{\partial \phi} + i\mathbf{e}_z \wedge, \\ &= -i\left(x\frac{\partial}{\partial y} - y\frac{\partial}{\partial x}\right) + i\mathbf{e}_z \wedge, \qquad (4.36) \\ &= L_z + S_z. \end{aligned}$$

Thus one part of the change produced by rotating a vector field is due to the variation of the field components at different field points. The differential operator L_z generates this part of the transformation. The other part of the change generated by S_z is due to a re-resolution of the vector field components when the field is rotated. The operator S_z can be written as a 3×3 matrix and is one of the infinitesimal transformation matrices of the representation $\mathscr{D}_1$ of the rotation group, thus the 'intrinsic spin' $S = 1$ properties associated with a vector field.

Let us define a set of three vector fields from the unit vectors $\mathbf{e}_x$, $\mathbf{e}_y$, and $\mathbf{e}_z$ by

$$\mathbf{e}_{\pm 1} = \mp\frac{1}{\sqrt{2}}(\mathbf{e}_x \pm i\mathbf{e}_y),$$

$$\mathbf{e}_0 = \mathbf{e}_z, \tag{4.37}$$

and three operators

$$S_{\pm} = (S_x \pm iS_y) = i(\mathbf{e}_x \pm i\mathbf{e}_y) \wedge,$$

$$S_z = i\mathbf{e}_z \wedge. \tag{4.38}$$

The operators (4.38) obey the commutation relations of angular momentum operators, and, acting on the set (4.37) of vector fields they transform them as components of a tensor of rank one, as may be proved by testing the relations (4.9). For example

$$S_{+}\mathbf{e}_0 = i(\mathbf{e}_x + i\mathbf{e}_y) \wedge \mathbf{e}_z = -(\mathbf{e}_x + i\mathbf{e}_y) = \sqrt{2}\,\mathbf{e}_1.$$

The field $\mathbf{r}$ provides an interesting example of the above analysis.

$$J_z\mathbf{r} = (L_z + S_z)(x\mathbf{e}_x + y\mathbf{e}_y + z\mathbf{e}_z),$$

$$= -i(-y\mathbf{e}_x + x\mathbf{e}_y) + i\mathbf{e}_z \wedge (x\mathbf{e}_x + y\mathbf{e}_y + z\mathbf{e}_z),$$

from the definition (4.36) of L_z and S_z.
Hence it follows that

$$J_z\mathbf{r} = 0,$$

a result which is not surprising because the field $\mathbf{r}$ is spherically symmetrical and a rotation transforms it into itself.

Vector fields which are generalizations of spherical harmonics may be formed by taking products of the vectors $\mathbf{e}_i$ defined in (4.37) with spherical harmonics and by using equation (4.6) to give an irreducible tensor.

$$\mathbf{Y}^{M}_{Ll1} = \sum_{mn} \langle LM|l1mn\rangle\, Y_{lm}\mathbf{e}_n. \tag{4.39}$$

The set of $2L+1$ fields $\mathbf{Y}^{M}_{Ll1}$ transform amongst themselves as components of a tensor rank L, i.e. according to equation (4.5). They are products of the tensors $\mathbf{e}_n$ and Y_{lm} of rank 1 and

l respectively. As an example the vector field $\mathbf{r}$ can be written as

$$\mathbf{r} = r\sqrt{\frac{4\pi}{3}}\sum_m (-1)^m Y_{1-m}\mathbf{e}_m$$
$$= -r\sqrt{4\pi}\,\mathbf{Y}^0_{011}(\theta\varphi).$$

Thus the field $\mathbf{r}$ is the scalar product of the spherical tensors rY_{1m} and $\mathbf{e}_m$ (both of rank 1) and is invariant for rotations as concluded above. In addition the vector spherical harmonics $Y^M_{Ll1}(\theta\varphi)$ satisfy the orthogonality conditions,

$$\int (\mathbf{Y}^M_{Ll1}(\theta\phi))^* . \mathbf{Y}^{M'}_{L'l'1}(\theta\varphi)\, d\Omega = \delta(ll')\,\delta(LL')\,\delta(MM'), \quad (4.40)$$

and they form a complete set for expanding the angular dependence of an arbitrary vector field.

It has been shown that the vector field $\mathbf{r}$ is invariant for rotations of axes. If ϕ_{LM} are a set of $2L+1$ scalar fields forming a spherical tensor of rank L (e.g. Y_{LM}) then the set of vector fields $\mathbf{r}\phi_{LM}$ also transforms as a spherical tensor of rank L. The vector field operators

$$\nabla = \mathbf{e}_x\frac{\partial}{\partial x}+\mathbf{e}_y\frac{\partial}{\partial y}+\mathbf{e}_z\frac{\partial}{\partial z},$$

$$\mathbf{L} = -i\mathbf{r}\wedge\nabla,$$

and

$$\nabla\wedge\mathbf{L},$$

share with the field $\mathbf{r}$ the property of being invariant under rotation. Hence each of the three sets of vector fields,

$$\nabla\phi_{LM},\ \mathbf{L}\phi_{LM} \quad\text{and}\quad \nabla\wedge\mathbf{L}\phi_{LM}, \quad (4.41)$$

form spherical tensors of rank L. In the special case when $\phi_{LM} = Y_{LM}$,

$$\mathbf{L}\,Y_{LM} = [L(L+1)]^{\frac{1}{2}}\,\mathbf{Y}^M_{LL1}$$

is a vector spherical harmonic. The fields (4.41) share with the vector spherical harmonics the property of orthogonality with respect to angular integrations and sometimes are a more convenient set than the vector spherical harmonics for making a multipole expansion of a vector field.

Such a case arises in considering solutions of the wave

equation for a vector field $\mathbf{A}$ with wave number k,

$$\nabla^2\mathbf{A}+k^2\mathbf{A} = 0. \tag{4.42}$$

The functions $\phi_{LM} = i^L(2L+1)j_L(kr)C_{LM}(\theta\phi)$ where $j_L(kr)$ are the spherical Bessel functions introduced in section 4.10.1 satisfy the scalar wave equation corresponding to equation (4.42), and with this choice of ϕ_{LM} the vector fields (4.41) form a complete set of expanding solutions of equation (4.42). In addition the set $\mathbf{A}_{LM}$, $\mathbf{A}^e_{LM}$, $\mathbf{A}^m_{LM}$ with

$$\begin{aligned} \mathbf{A}_{LM} &= (ik)^{-1}\nabla\phi_{LM} \\ \mathbf{A}^e_{LM} &= [k\sqrt{\{L(L+1)\}}]^{-1}\nabla\wedge\mathbf{L}\phi_{LM} \\ \mathbf{A}^m_{LM} &= [\sqrt{\{L(L+1)\}}]^{-1}\,\mathbf{L}\phi_{LM} \end{aligned} \tag{4.43}$$

have the same normalization as ϕ_{LM}. The fields $\mathbf{A}_{LM}$ are irrotational and have parity $(-1)^L$, while the fields $\mathbf{A}^e_{LM}$ and $\mathbf{A}^m_{LM}$ are solenoidal and have parity $(-1)^L$ and $(-1)^{L+1}$ respectively. The notation looks forward to the application of these fields in expanding the vector potential of the electromagnetic field. The fields $\mathbf{A}_{LM}$ are longitudinal fields whereas the fields $\mathbf{A}^e_{LM}$ and $\mathbf{A}^m_{LM}$ will represent the electric and magnetic components of the transverse field. It follows from equations (4.43) and (4.42) that

$$\mathbf{A}^e_{LM} = k^{-1}(\nabla\wedge\mathbf{A}^m_{LM}), \tag{4.44}$$

and

$$\mathbf{A}^m_{LM} = k^{-1}(\nabla\wedge\mathbf{A}^e_{LM}).$$

As an example of the expansion properties of the vector spherical harmonics and of the vector multipole fields (4.41) we obtain expansions of a plane wave in a vector field moving in the z-direction and with polarization vector $\mathbf{e}_q$. If $q = 0$ the wave has longitudinal polarization, if $q = \pm1$ the wave is transverse and has right or left circular polarization. Remembering the expansion of a scalar plane wave given in equation (4.30) and the definition (4.39) of vector spherical harmonics we get an equation

$$\begin{aligned} \mathbf{e}_q e^{ikz} &= \sum_l i^l\sqrt{\{4\pi(2l+1)\}}j_l(kr)\,Y_{l0}(\theta,\phi)\mathbf{e}_q, \\ &= \sum_{Ll} i^l\sqrt{\{4\pi(2l+1)\}}j_l(kr)\langle l10q|Lq\rangle\mathbf{Y}^q_{Ll1} \end{aligned} \tag{4.45}$$

as a series of vector spherical harmonics.

Alternatively we may expand the plane wave in terms of the fields (4.43). For the longitudinal component

$$\mathbf{e}_0 e^{ikz} = \frac{1}{ik}\nabla e^{ikz}$$

$$= \frac{1}{ik}\sum_L \nabla\phi_{L0}$$

$$= \sum_L \mathbf{A}_{L0}. \tag{4.46}$$

The transverse wave is a little more complicated. The symmetry about the z-axis determines that $\mathbf{e}_q e^{ikz}$ is a sum of components with $M = q = \pm 1$ and since the fields (4.43) form a complete set

$$\mathbf{e}_q e^{ikz} = \sum_L (a_L \mathbf{A}_{Lq} + b_L \mathbf{A}^m_{Lq} + c_L \mathbf{A}^e_{Lq}).$$

Taking the divergence of both sides of this equation determines that $a_L = 0$, taking the curl of both sides $(\nabla \wedge)$ and using equations (4.37) and (4.44) gives $b_L = qc_L$ and, finally, taking the scalar product of the operator $\mathbf{L}$ with both sides and using the expansion (4.30) for $\exp(ikz)$ gives $c_L = -1/\sqrt{2}$. Hence the transverse circularly polarized wave has expansion

$$\mathbf{e}_q e^{ikz} = -\frac{1}{\sqrt{2}}\sum_L (q\mathbf{A}^m_{Lq} + \mathbf{A}^e_{Lq}). \tag{4.47}$$

Equations (4.45) and (4.47) give the expansion of a plane polarized vector wave travelling in the direction of the z-axis. The expansions for a wave travelling in an arbitrary direction with wave number k can be obtained by rotating these expansions. For example the transverse wave has the expansion

$$\mathbf{e}_q e^{i\mathbf{k}.\mathbf{r}} = -\frac{1}{\sqrt{2}}\sum_{LM} (q\mathbf{A}^m_{LM} + \mathbf{A}^e_{LM})\mathscr{D}^L_{Mq}(R) \tag{4.48}$$

where R is the rotation taking the z-axis to the direction $\mathbf{k}$ (equation (2.22)). These calculations show that the $\mathbf{A}_{LM}$ are the natural fields for expanding a longitudinal vector wave. In the expansion of the vector potential in a transverse electromagnetic wave the components $\mathbf{A}^e_{LM}$ and $\mathbf{A}^m_{LM}$ correspond directly to its division into electric and magnetic multipoles.

4.10.3. Spinor Fields

The general spin-half field has two components and may be written

$$\mathbf{S} = \sum_\sigma \chi_\sigma S_\sigma(r, \theta, \phi),$$

where χ_σ is the two-component spinor,

$$\chi_{\frac{1}{2}} = \begin{pmatrix} 1 \\ 0 \end{pmatrix} \quad \text{and} \quad \chi_{-\frac{1}{2}} = \begin{pmatrix} 0 \\ 1 \end{pmatrix}.$$

The two components $S_{\frac{1}{2}}$ and $S_{-\frac{1}{2}}$ of $\mathbf{S}$ are defined with respect to a particular z-axis. A rotation changes the spatial dependence of the S_σ and also mixes the two components. Just as in the case of a vector field the infinitesimal rotation operator is a sum of two parts

$$J_i = L_i + S_i$$

where L_i acts on the spacial dependence of the field and S_i mixes its components. For the spin half-field the spin operators can be represented by the Pauli matrices $S_i = \frac{1}{2}\sigma_i$.

The analogues of the vector spherical harmonics are

$$\phi_{jln} = \sum_{\sigma m} \langle jn|\tfrac{1}{2}l\sigma m\rangle \chi_\sigma i^l C_{lm}(\theta\varphi).$$

Note that j is always half integral and can never be zero. Hence there is no spherically symmetric spinor field analogous to the vector field $\mathbf{r}$.

There is also an expansion of a spinor plane wave similar to (4.45)

$$\chi_\sigma e^{i\mathbf{k}.\mathbf{r}} = \sum_{ljn} (2l+1)\langle \tfrac{1}{2}l\sigma 0|j\sigma\rangle j_l(kr)\phi_{jln}\mathscr{D}^j_{n\sigma}(R).$$

The suffix σ refers to the polarization of the wave along $\mathbf{k}$, its direction of propagation.

4.10.4. *Electromagnetic Multipoles*

The electric and magnetic fields are given in terms of a vector potential **A** obeying the gauge condition div **A** $= 0$ by

$$\mathbf{H} = \text{curl}\,\mathbf{A},\ \mathbf{E} = -\frac{1}{c}\frac{\partial \mathbf{A}}{\partial t}.$$

The component of the vector potential which oscillates with wave number $k = \frac{\omega}{c}$ can be expanded in terms of the multipole fields (4.43) as

$$\begin{aligned}\mathbf{A} &= \mathbf{A}^e + \mathbf{A}^m \\ &= \sum_{LM} (q^e_{LM}\mathbf{A}^e_{LM} + q^m_{LM}\mathbf{A}^m_{LM}).\end{aligned} \qquad (4.49)$$

The fields $\nabla\phi_{LM}$ do not obey the gauge condition div **A** $= 0$, therefore do not appear in the expansion. The first terms of expression (4.49) denoted by $\mathbf{A}^e$ are the electric multipoles and the terms denoted by $\mathbf{A}^m$ are the magnetic multipoles.

On quantizing the electromagnetic field the Fourier coefficients q^e_{LM} and q^m_{LM} in (4.49) can be written in terms of creation and annihilation operators for photons (Heitler [34]).

If the photon states are normalized in a spherical box of radius R, then the electric multipole part of the vector potential takes the form

$$\mathbf{A}^e(\mathbf{r}) = \sum_{LM} \left(\frac{\hbar\omega}{R}\right)^{\frac{1}{2}} (a^e_{LM}\mathbf{A}^e_{LM}(\mathbf{r}) + a^{e+}_{LM}\mathbf{A}^{e+}_{LM}(\mathbf{r})) \qquad (4.50)$$

where a^{e+}_{LM} and a^e_{LM} are the creation and annihilation operators of electric 2^L-pole photons.‡ There is a similar expansion for the magnetic multipole part of the field.

‡ An electromagnetic transition probability is given by

$$q = \frac{2\pi}{\hbar}\left|\langle f\left|\frac{\mathbf{A}.\mathbf{p}}{\mu c}\right| i\rangle\right|^2 \rho(E)$$

where $\rho(E)$ is the density of final photon states. For waves of a definite angular momentum and for a spherical box of radius R, $\rho(E) = R/\pi\hbar c$. The radius of R of the region of quantization therefore cancels with the factor R in **A** (equation (4.50)).

4.10.5. Multipole Sources and Hyperfine Interactions

In potential problems where the extension of the source of a field may be neglected it is convenient to introduce the idea of a point source, for example a point charge or a point dipole in an electric field. Mathematically the source of the field produced by a unit point charge is represented by a δ-function‡ and the potential equation for the field $1/r$ is

$$\nabla^2 \frac{1}{r} = -4\pi\, \delta(\mathbf{r}).$$

If we differentiate both sides of this equation with respect to z we obtain the result that the source for the dipole field $\frac{z}{r^3}$ is $4\pi \frac{\partial \delta(\mathbf{r})}{\partial z}$.

This idea of a point source can be generalized by introducing a source function for the multipole field $\frac{1}{r^{L+1}} C_{LM}(\theta\phi)$ by the equation

$$\nabla^2 \left(\frac{1}{r^{L+1}} C_{LM}(\theta\phi) \right) = -4\pi\, \delta_{LM}(\mathbf{r}). \tag{4.51}$$

The source functions $\delta_{LM}(\mathbf{r})$§ are tensors of rank L, and are generalizations of the three-dimensional δ-function which vanish if $r \neq 0$ and have a singularity at $r = 0$ such that if $f(\mathbf{r})$ is an arbitrary scalar function regular at the origin

$$\int f^*(\mathbf{r})\, \delta_{LM}(\mathbf{r})\, d\mathbf{r} = F^*_{LM} \tag{4.52}$$

where F_{LM} is the coefficient of $r^L C_{LM}$ in a power series expansion for $f(\mathbf{r})$ about the origin.

In particular

$$\int r^{L'} C^*_{L'M'}\, \delta_{LM}(\mathbf{r})\, d\mathbf{r} = \delta(L'L)\, \delta(M'M).$$

‡ A δ-function is defined by the property that for any function $f(x)$ regular at $x = 0$

$$\int f(x)\, \delta(x)\, dx = f(0)$$

and the derivative of a δ-function $\delta'(x)$ by

$$\int \frac{d\delta(x)}{dx} f(x)\, dx = -\int \delta(x) \frac{df}{dx}\, dx = -f'(0),$$

a result obtained by a formal partial integration.

§ δ_{LM} is an Lth derivative of the 3-dimensional δ-function.

The set of singular functions may be further augmented by introducing functions

$$\delta^{n}_{LM} = \frac{(n+L)!\,(2L+1)!}{n!\,L!\,(2L+2n+1)!}\nabla^{2n}\,\delta_{LM}(\mathbf{r}) \tag{4.53}$$

and the resultant set includes all derivatives of the δ-function. The functions δ^{n}_{LM} have the property that for a scalar function $f(\mathbf{r})$

$$\int f^{*}(\mathbf{r})\,\delta^{n}_{LM}(\mathbf{r})\,d\mathbf{r} = F^{n*}_{LM} \tag{4.54}$$

where F^{n}_{LM} is the coefficient of $r^{L+2n}C_{LM}$ in a power series expansion of $f(\mathbf{r})$ about the origin. This result can be demonstrated by a series of formal partial integrations, reducing equation (4.54) to an integral of the form (4.52).

If we have an arbitrary scalar density distribution $\rho(\mathbf{r})$ concentrated near the origin we may introduce the formal series

$$\rho'(\mathbf{r}) = \sum_{nLM} Q^{n*}_{LM}\,\delta^{n}_{LM}(\mathbf{r}) \tag{4.55}$$

with coefficients Q^{n}_{LM} given by

$$Q^{n}_{LM} = \int r^{L+2n}C_{LM}(\theta,\phi)\rho(\mathbf{r})\,d\mathbf{r}. \tag{4.56}$$

Equation (4.55) can be considered as an expansion of the function $\rho(\mathbf{r})$ as a series of derivatives of the δ-function in the following sense. If $\phi(\mathbf{r})$ is a function of $\mathbf{r}$ which is expandable in a power series about the origin with a sphere of convergence which includes $\rho(\mathbf{r})$ then

$$\int \phi(\mathbf{r})\rho(\mathbf{r})\,d\mathbf{r} = \int \phi(\mathbf{r})\rho'(\mathbf{r})\,d\mathbf{r} \tag{4.57}$$

where $\rho'(\mathbf{r})$ is the series given in equation (4.55). The relation 4.57 may be proved by substituting the series for $\rho'(\mathbf{r})$ and using the property of the δ-functions given in equation (4.54). It may seem strange at first that the function $\rho(\mathbf{r})$ of finite extent may be represented as a series of δ-functions which take values only at the origin; but if it is remembered that equation (4.57) defines the meaning of the series $\rho'(\mathbf{r})$ and that $\phi(\mathbf{r})$ must be of finite extent then the difficulty is removed. The expression (4.55) may be used to find the potential $V(\mathbf{r})$ produced by a charge distribution $\rho(\mathbf{r})$. We have

$$\nabla^2 V = -4\pi\rho$$

and combining equations (4.51), (4.53), (4.55) we get

$$V(\mathbf{r}) = \sum_{LM} \left\{ Q^*_{LM} \frac{C_{LM}}{r^{L+1}} - \sum_{p=1}^{\infty} 4\pi Q^{p*}_{LM} \frac{\delta^{p-1}_{LM}(\mathbf{r})}{2n(2n+2L+1)} \right\}. \quad (4.58)$$

The first terms in (4.58) give the potential outside the charge distribution (equation (4.33)), while the further terms represent the modification of the potential inside the charge distribution.

The interaction energy between an extended charge distribution $\rho_e(\mathbf{r})$ and a concentrated distribution $\rho_n(\mathbf{r})$ (e.g. the potential energy of an electron in the field of a nucleus) can be found by expanding $\rho_n(\mathbf{r})$ as in equation (4.55) and obtaining $V_n(\mathbf{r})$ by equation (4.58). The interaction energy is

$$\begin{aligned} W &= \int \rho_e(\mathbf{r}) V_n(\mathbf{r})\, d\mathbf{r} \quad (4.59) \\ &= \sum_{LM} Q^*_{LM}(e) Q_{LM}(n), \end{aligned}$$

retaining only the first term in the expansion (4.58). The $Q_{LM}(n)$ are given by equation (4.56) and

$$Q_{LM}(e) = \int \frac{C_{LM}(\theta\phi)}{r^{L+1}} \rho_e(\mathbf{r})\, d\mathbf{r}. \quad (4.60)$$

Inclusion of higher terms in equation (4.58) gives an expansion of the interaction energy W as a power series in radius of the charge distribution $\rho_n(\mathbf{r})$.

A similar expansion can be made for the magnetic field produced by a static magnetization distribution $\mathbf{M}$ and a stationary current distribution of density $\mathbf{j}$. The magnetic induction $\mathbf{B}$ and the magnetic field $\mathbf{H}$ are given by

$$\mathbf{B} = \mathbf{H} + 4\pi\mathbf{M},$$

$$\text{curl}\ \mathbf{H} = 4\pi\mathbf{j},$$

$$\mathbf{B} = \text{curl}\ \mathbf{A},$$

and the vector potential $\mathbf{A}$ is chosen to satisfy the gauge condition div $\mathbf{A} = 0$. These equations yield an equation for the vector potential

$$\begin{aligned}\nabla^2\mathbf{A} &= -4\pi(\mathbf{j}+\text{curl }\mathbf{M}),\\ &= -4\pi\mathbf{J},\end{aligned}$$

introducing an effective current density $\mathbf{J}$ which is the sum of the charge current and the 'magnetization current'. Remembering the stationary current condition div $\mathbf{J} = 0$, $\mathbf{J}$ may be written in terms of two scalar fields as

$$\mathbf{J} = \mathbf{L}\phi_1+(\boldsymbol{\nabla}\wedge\mathbf{L})\phi_2.$$

If we retain only the lowest terms of a power series in the radius of the charge distribution it can be shown that the field ϕ_2 gives no contribution. Expanding ϕ_1 as a δ-function series (equation (4.55)) and keeping only the lowest terms for each multipole we can write

$$\mathbf{J} = \sum_{LM}\frac{1}{iL}M^*_{LM}\mathbf{L}\,\delta_{LM}(\mathbf{r}); \tag{4.61}$$

hence

$$\begin{aligned}\mathbf{L}\,.\,\mathbf{J} &= \sum_{LM}\frac{1}{iL}M^*_{LM}\mathbf{L}^2\,\delta_{LM}(\mathbf{r})\\ &= -i\sum_{LM}M^*_{LM}\,(L+1)\,\delta_{LM}(\mathbf{r}).\end{aligned}$$

Therefore the coefficients M_{LM} are given with the help of equation (4.56) as

$$M_{LM} = \frac{i}{L+1}\int r^L C_{LM}\,\mathbf{L}\,.\,\mathbf{J}\,d\mathbf{r}. \tag{4.62}$$

Since $\mathbf{J} = \mathbf{j}+\boldsymbol{\nabla}\wedge\mathbf{M}$ we see that the multipole moments M_{LM} have contributions from the current $\mathbf{j}$ and the magnetization $\mathbf{M}$. The current contribution has exactly the form of equation (4.62) with $\mathbf{j}$ replacing $\mathbf{J}$, while the magnetization part can be transformed by partial integration. Using the result of equation (6.6) (cf. Appendix VI) we get

$$M'_{LM} = \int\boldsymbol{\nabla}(r^L C_{LM})\,.\,\mathbf{M}\,d\mathbf{r}. \tag{4.63}$$

In a quantum system the current **j** is due to motion of charges while the magnetization **M** is associated with particle spins. For a single particle the current is represented by the operator

$$\mathbf{j} = \frac{e}{m} g_L\, \mathbf{p} = -i\frac{e}{m}\hbar g_L\, \nabla$$

and the magnetization by

$$\mathbf{M} = \frac{e\hbar}{2m} g_S \mathbf{S},$$

where m is the particle mass and g_L and g_S are its orbital and spin g-factors, (eg_L is the particle charge). In a many particle system **j** and **M** are sums of the corresponding single particle operators. With this identification the moments (4.62) and (4.63) are exactly equivalent to the dynamic magnetic moments (6.12b) and (6.12d).

Given the magnetic moments M_{LM} of the source the vector potential **A** can be calculated by using relations (4.51) and (4.61) giving

$$\mathbf{A}(\mathbf{r}) = \sum_{LM} \frac{i}{L} M^*_{LM} \mathbf{L}(r^{-(L+1)} C_{LM}(\theta, \phi)). \tag{4.64}$$

The induction **B** is found by evaluating curl **A** and making use of a result analogous to equation (6.6)

$$(\nabla \wedge \mathbf{L})(r^{-(L+1)} C_{LM}) = -iL\nabla(r^{-(L+1)} C_{LM}) + 4\pi i \mathbf{r}\, \delta_{LM}$$

giving

$$\mathbf{B} = \sum_{LM} M^*_{LM} \left[\nabla(r^{-(L+1)}\, C_{LM}) - \frac{4\pi}{L} \mathbf{r}\, \delta_{LM} \right]. \tag{4.65}$$

We shall show that the δ-function singularity in **B** at the origin is responsible for the spin contact interaction in hyperfine structure.

The Hamiltonian for the magnetic interaction between an electron and the nucleus of an atom is

$$W = \int (\mathbf{j}_e \cdot \mathbf{A}_n + \mathbf{M}_e \cdot \mathbf{B}_n)\, d\mathbf{r} \tag{4.66}$$

where $\mathbf{j}_e$ and $\mathbf{M}_e$ are operators representing the electron current and intrinsic magnetization and $\mathbf{A}_n$ and $\mathbf{B}_n$ are the vector potential and magnetic induction of the nuclear magnetic field. By replacing the fields by their multipole expansions we obtain

$$W = \sum_{LM} M^*_{LM}(e)\, M_{LM}(n). \tag{4.67}$$

The nuclear moments $M_{LM}(n)$ are given by equations (4.62) and (4.63) while $M_{LM}(e)$ is a sum of two parts, one being due to the electron orbital motion, the other to the spin interaction coming from the terms of equation (4.66) containing $\mathbf{j}_e$ and $\mathbf{M}_e$ respectively. The orbital part is (from equation (4.64))

$$M_{LM}(e) = \frac{i}{L}\int \mathbf{L}(r^{-(L+1)}C_{LM})\,.\,\mathbf{j}_e\, d\mathbf{r}. \tag{4.68}$$

and the spin part is (from equation (4.65))

$$M'_{LM}(e) = \int \nabla(r^{-(L+1)}C_{LM})\,.\,\mathbf{M}_e\, d\mathbf{r} - \frac{4\pi}{L}\int \mathbf{M}_e\,.\,\mathbf{r}\;\delta_{LM}\, d\mathbf{r}. \tag{4.69}$$

For $L = 1$ the last term in equation (4.69) is the familiar contact interaction between the electron spin and the nuclear spin when the electron is in an S-orbit. These are the same as the formulae for multipole interactions given by Schwartz [59].

Equation (4.66) represents the magnetic interaction between a classical electron and the nuclear magnetic field. The corresponding relativistic interaction is

$$W = \int e\boldsymbol{\alpha}\,.\,\mathbf{A}_n\, d\mathbf{r}, \tag{4.70}$$

where $\boldsymbol{\alpha}$ is the Dirac [20] matrix giving the electron velocity. The multipole expansions still hold if $\mathbf{j}_e$ is replaced by $e\boldsymbol{\alpha}$ and $\mathbf{M}_e \to 0$ because the spin interaction is already included in (4.70).

CHAPTER V

MATRIX ELEMENTS OF TENSOR OPERATORS

WE are now in a position to calculate those parts of any physical problem which are concerned with angular integrations, the coupling of angular momentum vectors, or our choice of coordinate system—in fact general rotational symmetries—which we might call the 'geometrical' parts. The basic technique is to exploit the simplest matrix reduction which we have already met in Chapter IV, namely the Wigner–Eckart theorem for the matrix elements of spherical tensors

$$\langle JM|T_{KQ}|J'M'\rangle = (-1)^{2K}\langle JM|J'KM'Q\rangle\langle J||\mathbf{T}_K||J'\rangle. \quad (5.1)$$

The aim is to derive some general formulae for matrix elements which depend only on the geometrical structure of the states and tensors involved, so then they can be applied to a variety of physical situations. Some of these applications are described in the next chapter. The results of such calculations are expressed in terms of the algebraic functions discussed in Chapter III. Owing to the extensive tabulation of these functions for a wide range of values of their arguments, this is a form very suitable for numerical work.

5.1. Projection Theorem for Vector Operators

First, we draw attention to the so-called Projection Theorem for vector operators $\mathbf{V}$, which is the special case of the Wigner-Eckart theorem for tensors of rank one. It concerns matrix elements between states with the same J,

$$\begin{aligned}\langle JM'|\mathbf{V}|JM\rangle &= \langle JM'|\mathbf{J}(\mathbf{J}\,.\,\mathbf{V})|JM\rangle/J(J+1)\\ &= g_J(V)\langle JM'|\mathbf{J}|JM\rangle. \end{aligned} \quad (5.2)$$

This may be proved by expanding the right hand side, with the help of the Wigner–Eckart theorem. The physical significance of (5.2) becomes obvious when we realize that

$$\mathbf{J}(\mathbf{J}\,.\,\mathbf{V})/J(J+1)$$

is the component of $\mathbf{V}$ along the unit vector

$$\mathbf{J}/\sqrt{\{J(J+1)\}}.$$

According to the vector model, $\mathbf{V}$ is precessing about $\mathbf{J}$, and its component perpendicular to $\mathbf{J}$ averages to zero (cf. Fig. 5). The theorem states that the expectation value of a vector operator in a state of sharp $\mathbf{J}$ is always proportional to the expectation value of $\mathbf{J}$, as the second form of (5.2) emphasis. The coefficient $g_J(\mathbf{V})$ is a generalization of the g-factor familiar in magnetic moment problems. In reduced form, (5.2) becomes

$$\langle J||\mathbf{V}||J\rangle = g_J(\mathbf{V})\sqrt{\{J(J+1)\}}. \tag{5.3}$$

5.2. Matrix Elements of Tensor Products

Next we consider a tensor $T_{KQ}(k_1k_2)$ which is itself a tensor product of tensors of rank k_1 and k_2, as described in Chapter IV.

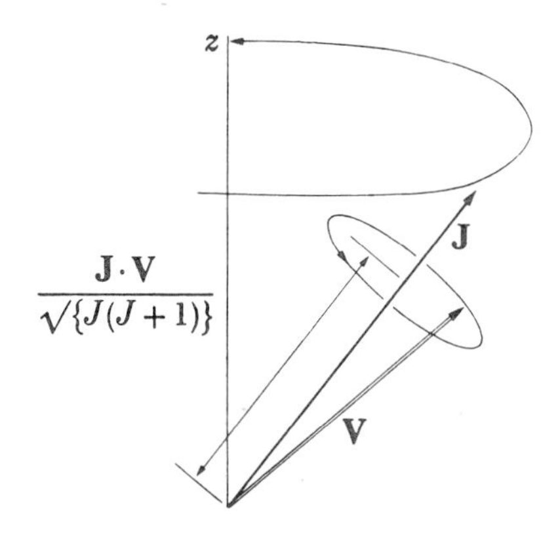

Fig. 5. The vector $\mathbf{V}$ precesses rapidly about $\mathbf{J}$ so that its component at right angles to $\mathbf{J}$ averages to zero. Then the component of $\mathbf{V}$ along O_z is determined by the direction of $\mathbf{J}$.

We may write such a tensor

$$T_{KQ}(k_1k_2) = \sum_q R_{k_1q}S_{k_2Q-q}\langle k_1k_2qQ-q|KQ\rangle. \tag{5.4}$$

The reduced matrix of the composit tensor $\mathbf{T}$ may be evaluated in terms of the matrices of the $\mathbf{R}$ and $\mathbf{S}$ by using (5.4) and introducing intermediate states between the two tensors,

$$\langle J||\mathbf{T}_K||J'\rangle = \sum_Q (-1)^{2K}\langle JM|T_{KQ}|J'M-Q\rangle\langle JM|J'KM-QQ\rangle,$$

$$= \sum_{QqJ''} (-1)^{2K}\{\langle JM|J'KM-QQ\rangle\langle KQ|k_1k_2qQ-q\rangle \times$$

$$\times \langle JM|R_{k_1q}|J''M-q\rangle\langle J''M-q|S_{k_2Q-q}|J'M-Q\rangle\},$$

$$= \sum_{J''} (-1)^{K-k_1-k_2}\{(2J''+1)(2K+1)\}^{\frac{1}{2}}\, W(JJ'k_1k_2;\, KJ'') \times$$

$$\times \langle J||\mathbf{R}_{k_1}||J''\rangle\langle J''||\mathbf{S}_{k_2}||J'\rangle. \tag{5.5}$$

The geometrical significance of the Racah function is readily understood when we notice we have carried out a recoupling of three angular momentum vectors, expressing a scheme like $[(k_1k_2)K,J';J]$ in terms of one like $[k_1,(k_2J')J'';J]$. The Racah W is just the transformation coefficient between these two schemes.

When $K = 0$, (5.4) defines the scalar product of two tensors, although a different normalization is convenient (cf. equation (4.7))

$$\mathbf{R}_k \cdot \mathbf{S}_k = (-1)^k\sqrt{(2k+1)}T_{00}(kk). \tag{5.6}$$

Equation (5.5) then reduces to

$$\langle J||\mathbf{R}_k \cdot \mathbf{S}_k||J\rangle = \sum_{J'} (-1)^{-J+J'}\left(\frac{2J'+1}{2J+1}\right)^{\frac{1}{2}} \langle J||\mathbf{R}_k||J'\rangle\langle J'||\mathbf{S}_k||J\rangle. \tag{5.7}$$

If $\mathbf{S}_k = \mathbf{R}_k$, and is Hermitian in the sense of (4.23), we may use (4.24) to reduce (5.7) further

$$\langle J||\mathbf{R}_k \cdot \mathbf{R}_k||J\rangle = \sum_{J'} (-1)^p|\langle J||R||J'\rangle|^2. \tag{5.8}$$

We have said nothing about the nature of the tensors $\mathbf{R}_k$ and $\mathbf{S}_k$; in the next section we shall consider the special case of a system composed of two parts, with $\mathbf{R}$ acting on one, $\mathbf{S}$ on the other. In such a case (5.5) can be expanded further.

5.3. Reduction of Matrices for Composite Systems

We are often faced with the problem of computing the value of an operator which refers to only one part of a composite system. Such a composite system may consist of two or more separate systems, or merely different aspects of the same system, such as the spin and orbital properties. Suppose we have a two-component system with angular momenta j_1 and j_2, and resultant J, while the tensor $T_k(1)$ acts only on part 1 of the system. The coupled state vectors are given by (2.27), and the reduced matrix of $T_k(1)$ found by expanding them

$$\langle j_1 j_2 J \| \mathbf{T}_k(1) \| j_1' j_2' J' \rangle = \sum_q (-1)^{2k} \langle JM | J'kM-qq \rangle \times \\ \times \langle j_1 j_2 JM | T_{kq}(1) | j_1' j_2' J'M-q \rangle$$

$$= \sum_{qm} \langle JM | J'kM-qq \rangle \langle j_1 j_2 mM-m | JM \rangle \times \\ \times \langle j_1' j_2' m-qM-m | J'M-q \rangle \times \\ \times \langle j_1 m | j_1' km-qq \rangle \langle j_1 \| \mathbf{T}_k(1) \| j_1' \rangle \, \delta(j_2 j_2'),$$

$$= (-1)^{J+j_1'-k-j_2} \{(2J'+1)(2j_1+1)\}^{\frac{1}{2}} \, W(j_1 j_1' J J'; k j_2) \times \\ \times \langle j_1 \| \mathbf{T}_k(1) \| j_1' \rangle \, \delta(j_2 j_2'). \qquad (5.9)$$

The Racah function appears for precisely the same reason as it does in (5.5). The mirror formula to (5.9), when $\mathbf{T}_k(2)$ acts only on part 2 of the system, is easily obtained by remembering that changing the order of coupling of parts 1 and 2 only introduces a phase,

$$|j_1 j_2 JM\rangle = (-1)^{j_1+j_2-J} |j_2 j_1 JM\rangle,$$

from the symmetry of the Clebsch–Gordan coefficient.

A simple example is that of a particle with spin, and an operator which acts only on the spatial coordinates. Such an operator could be the spherical harmonic C_{LM}. If the particle's total angular momentum j is made up of orbital l and spin s,

$$\langle lsj \| \mathbf{C}_L \| l'sj' \rangle = (-1)^{j-L-s+l'} [(2l+1)(2j'+1)]^{\frac{1}{2}} \times \\ \times W(ll'jj'; Ls) \langle l \| \mathbf{C}_L \| l' \rangle \quad (5.10)$$

where $\langle l \| \mathbf{C}_L \| l' \rangle$ has been given in (4.17.) When $l+l'+L$ is even and $s = \frac{1}{2}$ this simplifies considerably when the relation for W in Appendix II is used,

$$\langle l\tfrac{1}{2}j \| \mathbf{C}_L \| l'\tfrac{1}{2}j' \rangle = (-1)^{j'-\frac{1}{2}+L} \sqrt{(2j'+1)} \begin{pmatrix} j & j' & L \\ \frac{1}{2} & -\frac{1}{2} & 0 \end{pmatrix}. \quad (5.11)$$

which is independent of l and l'.

More generally we may have a product of tensors with $\mathbf{R}(1)$ acting on part 1 and $\mathbf{S}(2)$ on part 2 of our composite system. The matrix of such an operator may be expressed in terms of the matrices of the component systems in the same way as (5.9), or by noticing that in this case each of the matrix elements on the right hand of (5.5) are of the form (5.9). Either way we have to change the coupling of the four vectors j_i from a scheme $[(j_1j_2)J, (j_1'j_2')J'; K]$ to one like $[(j_1j_1')k_1(j_2j_2')k_2; K]$ which indicates that an X-coefficient is involved. We soon find

$$\langle j_1j_2J||\mathbf{T}_K(k_1k_2)||j_1'j_2'J'\rangle = [(2J'+1)(2K+1)(2j_1+1)\times$$
$$\times\ (2j_2+1)]^{\frac{1}{2}} \times \begin{Bmatrix} J & J' & K \\ j_1 & j_1' & k_1 \\ j_2 & j_2' & k_2 \end{Bmatrix} \langle j_1||\mathbf{R}_{k_1}||j_1'\rangle\langle j_2||\mathbf{S}_{k_2}||j_2'\rangle. \quad (5.12)$$

Putting $k_2 = 0$ of course, (5.12) reduces to (5.9). When $K = 0$, we have the scalar product (5.6), and (5.12) becomes

$$\langle j_1j_2J||\mathbf{R}_k(1)\,.\,\mathbf{S}_k(2)||j_1'j_2'J'\rangle = \{(2j_1+1)(2j_2+1)\}^{\frac{1}{2}}\times$$
$$\times\ \delta(J,J')(-1)^{J-j_1-j_2'}\,W(j_1j_1'j_2j_2';kJ)\langle j_1||\mathbf{R}_k||j_1'\rangle\langle j_2||\mathbf{S}_k||j_2'\rangle. \quad (5.13)$$

(This may also be obtained from (5.7) by using the Racah sum rule (3.21).)

There is a simple geometric interpretation of the diagonal elements of (5.13) with $j_1 = j_1', j_2 = j_2'$. The reduced elements on the right hand are the average or expectation values of **R** along $\mathbf{j}_1$ and **S** along $\mathbf{j}_2$, respectively (see 5.2), while in the limit of large j_1, j_2 and J [Biedenharn 5],

$$(-1)^{J-j_1-j_2}[(2j_1+1)(2j_2+1)]^{\frac{1}{2}}\,W(j_1j_1j_2j_2,kJ) \simeq P_k(\cos\theta).$$

Here θ is the angle between the vectors $\mathbf{j}_1$ and $\mathbf{j}_2$.

$$\cos\theta = \{J(J+1)-j_1(j_1+1)-j_2(j_2+1)\}/2j_1j_2.$$

For $k = 1$, this is the familiar form for the scalar product of two vectors.

There are many such scalar product operators, one example being the scalar product of the renormalized spherical harmonics of (2.9).

$$\langle l_1l_2L||\mathbf{C}_k(1)\,.\,\mathbf{C}_k(2)||l_1'l_2'L'\rangle$$
$$= \delta(L',L)(-1)^{L+k}[(2l_1+1)(2l_2+1)(2l_1'+1)(2l_2'+1)]^{\frac{1}{2}}\times$$
$$\times\,W(l_1l_1'l_2l_2';kL)\begin{pmatrix} k & l_1 & l_1' \\ 0 & 0 & 0 \end{pmatrix}\begin{pmatrix} k & l_2 & l_2' \\ 0 & 0 & 0 \end{pmatrix}. \quad (5.14)$$

Another important example is the spin-orbit coupling operator $\mathbf{l}\,.\,\mathbf{s}$, with $k = 1$. This has matrix elements

$$\langle lsj||\mathbf{l}\,.\,\mathbf{s}||l's'j'\rangle = \delta(s,s')\,\delta(l,l')\,\delta(j,j')(-1)^{j-l-s}W(llss,1j)\times$$
$$\times[l(l+1)(2l+1)\,s(s+1)(2s+1)]^{\frac{1}{2}}.$$

The explicit form for the Racah function form Table 4 leads to

$$\langle lsj||\mathbf{l}\,.\,\mathbf{s}||lsj\rangle = \tfrac{1}{2}[j(j+1)-l(l+1)-s(s+1)]. \qquad (5.15)$$

In this particular case, of course, this result could have been obtained far more easily by using the vector relation

$$2\mathbf{l}\,.\,\mathbf{s} = \mathbf{j}^2-\mathbf{l}^2-\mathbf{s}^2. \qquad (5.16)$$

It should perhaps be emphasized that in the explicit examples (5.11), (5.14), and (5.15) above we are concerned only with angular and spin integration; any radial integration has not been included explicitly.

Other more complex systems with more than two components may be dealt with by repeated application of the techniques described here, or by the methods described in the next section.

5.4. Systems of Many Particles

When dealing with systems of n similar particles, we are interested mainly in the matrix elements of two simple types of operators. The first is the one-body operator $F = \sum_{i=1}^{n} f(i)$, where $f(i)$ acts only on the coordinates of the ith particle. The other is the two-body operator $G = \sum_{i>j} g(ij)$, where

$$g(ij) = g(ji)$$

acts only between particles i and j, and $i > j$ means a sum over all pairs.

It is not appropriate here to go into the details of the construction of multi-particle wave functions; the reader is referred to other standard works [14], [17], [26], [49]. We shall merely indicate briefly the technique for evaluating matrix elements of operators like F and G.

The wave function for a number n of identical fermions

(bosons) must be written so that it is antisymmetric (symmetric) under the exchange of the coordinates of any two particles. Thus it is not possible in general to write it as the simple product of an antisymmetric (symmetric) function for $n-1$ of the particles times a wave function for the nth. However, it must be possible to write it as a suitable sum of such products. Suppose the system had total angular momentum J and z-component M, we may write (summed over α_p, J_p, a and j)

$$|\alpha(n), JM\rangle = \sum |\alpha_p(n-1)J_p, aj; JM\rangle\langle\alpha_p(n-1)J_p, aj|\}\alpha(n)J\rangle, \tag{5.17}$$

where $$|\alpha_p(n-1)J_p, aj; JM\rangle = \sum_m |\alpha_p(n-1)J_p\, M-m\rangle|ajm\rangle_n \times \times\langle J_p\, jM-mm|JM\rangle, \tag{5.18}$$

and α, α_p, a are any additional quantum numbers needed to specify the states fully. The parent states $|\alpha_p J_p\, M_p\rangle$ are orthonormal and fully antisymmetric (symmetric) in the first $n-1$ particles, while the nth occupies the state $|ajm\rangle_n$. The individual terms of (5.17) are not fully antisymmetric (symmetric) in all n particles, but their sum, weighted by the numerical coefficients of fractional parentage (cfp)

$$\langle\alpha_p J_p, aj|\}\alpha J\rangle$$

must be. The cfp describe how the state $|\alpha JM\rangle$ may be built up from its possible parent states obtained by the removal of one particle. The second relation (5.18) merely describes the vector coupling of the two parts to the correct total angular momentum. The orthonormality of the various states in the expansion (5.17) ensures that the cfp obey

$$\sum_{\alpha_p J_p aj} \langle\alpha J\{|\alpha_p J_p\, aj\rangle\langle\alpha_p J_p\, aj|\}\alpha' J'\rangle = \delta(\alpha, \alpha')\,\delta(J, J'). \tag{5.19}$$

The expansion 5.17 is of principal interest when the many-body wave function $|\alpha JM\rangle$ describes a single 'independent particle' configuration, that is, one in which each particle may be assigned a set of quantum numbers (aj) (an 'orbit') independently of the others, the whole suitably vector-coupled to give the correct total angular momentum. Configurations of a number of 'equivalent' fermions, that is, all occupying the same orbit (aj), have been widely studied in the literature and many of the cfp tabulated [26].

Since the n similar particles are indistinguishable, the matrix element of the one-body operator F must be just n times that of $f(n)$. Suppose F were a tensor of rank k, $\mathbf{F}_k$. When we use the wave function (5.17), $\mathbf{f}_k(n)$ only acts on the nth particle, so we have a sum of matrix elements of the form (5.9) discussed above. We find

$$\begin{aligned}&\langle\alpha(n)J||\mathbf{F}_k||\alpha'(n)J'\rangle\\&\quad= n\sum(-1)^{J_p+k-J-j'}\{(2J'+1)(2j+1)\}^{\frac{1}{2}}W(jj'JJ';kJ_p)\times\\&\quad\times\langle aj||\mathbf{f}_k||a'j'\rangle\langle\alpha(n)J\{|\alpha_p(n-1)J_p,aj\rangle\langle\alpha_p(n-1)J_p\,a'j'|\}\alpha'(n)J'\rangle\end{aligned} \tag{5.20}$$

summed over α_p, J_p, a, a', j, j'. Because of the orthogonality of the parent wave functions, only those parent states common to $|\alpha(n)JM\rangle$ and $|\alpha'(n)J'M\rangle$ can contribute to (5.20). A very interesting result appears when we have n equivalent particles, all with angular momentum j, so the sum over j, j' reduces to one term and

$$\begin{aligned}\langle\alpha(j^n)J||\mathbf{F}_k||\alpha'(j^n)J'\rangle &= n\langle aj||\mathbf{f}_k(n)||aj\rangle\times\\&\times\sum_{\alpha_pJ_p}[(-1)^{J_p+k-J-j}\{(2J'+1)(2j+1)\}^{\frac{1}{2}}W(jjJJ';kJ_p)\times\\&\times\langle\alpha J\{|\alpha_pJ_p,aj\rangle\langle\alpha_pJ_p,aj|\}\alpha'J'\rangle].\end{aligned} \tag{5.21}$$

When F is scalar, $k = 0$, this becomes

$$\langle\alpha(j^n)J||\mathbf{F}_0||\alpha'(j^n)J'\rangle = n\,\delta(\alpha,\alpha')\,\delta(J,J')\langle aj||\mathbf{f}_0||aj\rangle. \tag{5.22}$$

The form of (5.21) shows that for these equivalent particle configurations the ratio of the matrix elements of two such one-body operators, $\mathbf{F}_k$ and $\mathbf{R}_k$, is equal to the ratio of the corresponding single particle matrix elements.

$$\frac{\langle\alpha(j^n)J||F_k||\alpha'(j^n)J'\rangle}{\langle\alpha(j^n)J||R_k||\alpha'(j^n)J'\rangle} = \frac{\langle aj||f_k||aj\rangle}{\langle aj||r_k||aj\rangle}, \tag{5.23}$$

independently of α, α', J and J'. In particular, with the vector operator $\mathbf{V}$ we may take $\mathbf{R}_1$ to be $\mathbf{J}$, obtaining the simple result

$$\langle\alpha(j^n)J||\mathbf{V}||\alpha'(j^n)J'\rangle = \delta(\alpha,\alpha')\,\delta(J,J')\left(\frac{J(J+1)}{j(j+1)}\right)^{\frac{1}{2}}\langle aj||\mathbf{v}||aj\rangle. \tag{5.24}$$

When discussing the projection theorem for vector operators in section 5.1 we defined a generalized g-factor by (5.2), (5.3). The relation (5.24) shows that for one-body operators the g-factor for n equivalent particles is the same as for one particle, independent of J and α. Putting $F_{k.} = J$ in equation (5.21) yields a sum rule for cfp.

Matrix elements of the two-body operator G may be evaluated in the same way, except that now we need to apply the fractional parentage expansion twice in order to 'peel off' the nth and $(n-1)$th particles. The value of G is then the value of $g(n, n-1)$, times the number of pairs, $\frac{1}{2}n(n-1)$. In practice the two steps are often combined, and cfp defined for the reduction of the n-particle wave function to products of a function for the first $n-2$ times one for the last pair, appropriately symmetrized [26]. Again, only parent states of the $n-2$ particles common to both n particle states can contribute to matrix elements of G between the latter. With independent particle wave functions this means the configurations for the two states can differ at most in two sets of single particle quantum numbers if the matrix element is not to vanish.

5.5. Isotopic Spin Formalism

In nuclear problems, neutrons and protons are often treated on the same footing by introducing isotopic (or 'isobaric') spin.

The i-spin operator $\boldsymbol{\tau}$ is identical with the Pauli spin operator $\boldsymbol{\sigma}$, except that it is said to operate in an abstract charge, or *i-spin*, space. The eigenvalues ± 1 of $\boldsymbol{\tau}_z$ are then used to denote neutron or proton respectively. Charge independence of nuclear forces then leads to conservation of total *i*-spin, $\mathbf{T} = \sum_i \boldsymbol{\tau}_i$, and nuclear state vectors are simple products of ordinary space-spin functions with an *i*-spin function, $|\alpha JM, \alpha_I TM_I\rangle = |\alpha JM\rangle|\alpha_I TM_I\rangle$. Since the operators $\boldsymbol{\tau}$ and $\mathbf{T}$ behave in charge space just as angular momenta in ordinary space, all the vector coupling and other techniques described in this book apply equally to them and their eigenfunctions. In particular, we may construct operators in this space (which, just as in Pauli spin space, can always be expressed in terms of $\boldsymbol{\tau}$ and the unit operator 1), whose matrix elements are to be evaluated as described in this chapter. If the interactions are charge-independent the operators representing them must be scalar in the product charge space for the whole system, but if not (such as for electromagnetic interactions) higher rank tensors will appear. For example, the two-body charge exchange operator P^τ can be written in the scalar form

$$P^\tau_{12} = \tfrac{1}{2}(1+\boldsymbol{\tau}_1 \cdot \boldsymbol{\tau}_2),$$

while the electrostatic interaction of two nucleons take the form

$$(1-\tau_{1z})(1-\tau_{2z})\frac{e^2}{4r_{12}}$$
$$= [1-\tau_{1z}-\tau_{2z}+\tfrac{1}{3}(\boldsymbol{\tau}_1 \cdot \boldsymbol{\tau}_2)+\sqrt{\tfrac{2}{3}}T_{20}(\tau_1, \tau_2)]\, e^2/4r_{12},$$

in which scalar, vector, and second rank tensors appear.

CHAPTER VI

APPLICATIONS TO PHYSICAL SYSTEMS

THE purpose of the present chapter is to apply the results of the previous two chapters to some physical systems of interest in atomic and nuclear physics, as examples of the usefulness of these techniques.

6.1. Electromagnetic Radiative Transitions

In this section we shall use the multipole expansion of a vector field, described in section 4.10.2, to define the electromagnetic interaction tensors and to calculate their matrix elements for the emission of a photon. We confine ourselves to non-relativistic particles, which is usually sufficient for nuclear problems. The treatment of relativistic motion follows very similar lines [34], [53], [54].

A particle of spin **s** and momentum **p** moving in an electromagnetic field whose vector potential is **A** experiences (in the Lorentz gauge) an interaction [59],

$$H'(\mathbf{r}) = -\frac{e}{2mc}\{g_L[\mathbf{A}(\mathbf{r})\,.\,\mathbf{p}+\mathbf{p}\,.\,\mathbf{A}(\mathbf{r})]+g_s\hbar\mathbf{s}\,.\,\nabla\wedge\mathbf{A}(\mathbf{r})\}. \tag{6.1}$$

g_s is the spin g-factor so that the intrinsic magnetic moment of the particle is $\boldsymbol{\mu} = g_s\mathbf{s}(e\hbar/2mc)$, while g_L is its orbital g-factor and eg_L its charge. Terms quadratic in **A** are neglected in (6.1) since they lead to two-photon emission. Also, the first two terms of (6.1) are identical since **A** commutes with **p** in this gauge (div $\mathbf{A} = 0$).

6.1.1. *The Multipole Operators*

The multipole expansion of the vector potential $\mathbf{A}_q$ for a circularly polarized plane wave has already been described in section 4.10.2. It remains to insert these multipoles into (6.1) in order to find the corresponding interaction tensors. From (4.48) we have

$$\mathbf{A}_q(\mathbf{k},\mathbf{r}) = \mathbf{e}_q e^{i\mathbf{k}.\mathbf{r}} = -\frac{1}{\sqrt{2}}\sum_{LM}(q\mathbf{A}^m_{LM}(\mathbf{r})+\mathbf{A}^e_{LM}(\mathbf{r}))\mathscr{D}^L_{Mq}(R) \tag{6.2}$$

where $\quad \sqrt{\{L(L+1)\}}\mathbf{A}^m_{LM} = i^L(2L+1)j_L(kr)\mathbf{L}C_{LM}$

and

$$\mathbf{A}^e_{LM} = (1/k)\nabla\wedge\mathbf{A}^m_{LM}, \tag{6.3}$$

corresponding to a flux of $(k/4\pi\hbar)$ photons per second along **k**. R is the rotation taking the quantization axes into those with z-axis along **k**, and $q = \pm 1$ for left/right circular polarization about **k**.

The charge dependent terms of (6.1) are proportional to $\mathbf{A}\,.\,\boldsymbol{\nabla}$. The magnetic potential of (6.3) can be used as it stands, but it is more convenient to use the vector relation which follows from $\mathbf{L}\,.\,\boldsymbol{\nabla} = 0$,

$$\mathbf{L}f(\mathbf{r})\,.\,\boldsymbol{\nabla} = -\boldsymbol{\nabla}f(\mathbf{r})\,.\,\mathbf{L}, \tag{6.4}$$

because the operator $\mathbf{L}$ commutes with the radial part of the particle wave function while $\boldsymbol{\nabla}$ does not. This is particularly advantageous when the long wavelength approximation is used,

$$j_L(kr) \simeq (kr)^L/(2L+1)!!, \quad \text{if} \quad kr \ll 1 \tag{6.5}$$

Then $f(\mathbf{r})$ is proportional to the solid harmonic $r^L C_{LM}$, and the gradient formula of Appendix VI can be used.

The long wavelength approximation also simplifies the electric radiation field in (6.3), for we have the relation

$$\boldsymbol{\nabla} \wedge \mathbf{L}(r^L C_{LM}) = i(L+1)\boldsymbol{\nabla}(r^L C_{LM}) \tag{6.6}$$

and the electric part of the vector potential is the gradient of a scalar potential. We then use the anti-hermitian property of $\boldsymbol{\nabla}$ and the Schrödinger equation for the states $|\alpha\rangle$ of the radiating particle,

$$[-(\hbar^2/2m)\boldsymbol{\nabla}^2 + V - E_\alpha]|\alpha\rangle = 0, \tag{6.7}$$

to simplify the electric radiation matrix element. With (6.6) the electric component of the vector potential (6.3) is proportional to $\boldsymbol{\nabla}f(\mathbf{r})$ where $f(\mathbf{r})$ is the solid harmonic $r^L C_{LM}$. The matrix element of this part of (6.1) is then proportional to

$$\begin{aligned}
&\langle\beta|\boldsymbol{\nabla}f\,.\,\boldsymbol{\nabla} + \boldsymbol{\nabla}\,.\,\boldsymbol{\nabla}f|\alpha\rangle \\
&= \langle\beta|\boldsymbol{\nabla}f\,.\,\boldsymbol{\nabla}|\alpha\rangle - \langle\alpha|\boldsymbol{\nabla}f\,.\,\boldsymbol{\nabla}|\beta\rangle^* \\
&= -\langle\beta|f\boldsymbol{\nabla}^2|\alpha\rangle + \langle\alpha|f\boldsymbol{\nabla}^2|\beta\rangle^* \\
&= (E_\alpha - E_\beta)(2m/\hbar^2)\langle\beta|f|\alpha\rangle.
\end{aligned} \tag{6.8}$$

We have assumed the potential V in (6.7) is hermitian. Thus the effective electric multipole transition operator for long wavelengths is the solid harmonic $r^L C_{LM}$, which is the same as the operator for the electrostatic moments, Q_{LM}, derived in section 4.10.1. This result is easily shown to hold for systems of more than one particle (the energies E_α, E_β are then the total energies of the system). A derivation parallel to (6.8) uses the continuity equation [9].

For the spin terms of the interaction (6.1) we require the magnetic field vector $\mathbf{H} = \nabla \wedge \mathbf{A}$. From (6.3) we see immediately that

$$\mathbf{H}^m_{LM} = k\mathbf{A}^e_{LM}. \tag{6.9}$$

An analogous result follows for $\mathbf{H}^e$ from (6.3) if we remember the wave equation for $\mathbf{A}$, $\nabla \wedge (\nabla \wedge \mathbf{A}) = k^2\mathbf{A}$,

$$\mathbf{H}^e_{LM} = k\mathbf{A}^m_{LM}. \tag{6.10}$$

Collecting these results together we can write down the multipole tensor expansion of the interaction (6.1) when $\mathbf{A}_q$ represents a plane wave circularly polarized about its direction of motion.

$$H'_q(\mathbf{k}, \mathbf{r}) = -\sum_{LM} \frac{i^L k^L}{(2L-1)!!} \sqrt{\frac{L+1}{2L}} \mathscr{D}^L_{Mq}(R) \times$$
$$\times \{(Q_{LM}+Q'_{LM}) - iq(M_{LM}+M'_{LM})\}. \tag{6.11}$$

Q_L and Q'_L are used to denote, respectively, the charge and spin contributions of parity $(-)^L$, leading to electric radiation, while similarly M_L and M'_L have parity $(-)^{L+1}$ and lead to magnetic radiation. Explicitly, in the long wavelength approximation ($kr \ll 1$), and with the magneton $\beta = e\hbar/2mc$,

$$Q_{LM}(\mathbf{r}) = eg_L(r^L C_{LM}), \tag{6.12a}$$

$$M_{LM}(\mathbf{r}) = 2\beta g_L \nabla(\mathbf{r}^L C_{LM}) \,.\, \mathbf{L}/(L+1), \tag{6.12b}$$

$$Q'_{LM}(\mathbf{r}) = -k\beta g_s \mathbf{L}(r^L C_{LM}) \,.\, \mathbf{S}/(L+1), \tag{6.12c}$$

$$M'_{LM}(\mathbf{r}) = \beta g_s \nabla(r^L C_{LM}) \,.\, \mathbf{S}. \tag{6.12d}$$

These spherical tensors are of quite general application;‡ the coefficients outside the brackets in (6.11) merely select the

‡ The multipole tensors (6.12) agree with those used by some authors [11], [50], [59]. They are $[4\pi/(2L+1)]^{\frac{1}{2}}$ times the tensors defined by others [9], [13], [63], because of the use in (6.12) of the more convenient C_{LM} instead of Y_{LM}. In addition to this, the Blatt and Weisskopf [9] multipole tensors are the hermitian adjoints of ours.

combination appropriate to the emission or absorption of a photon with definite linear momentum **k** and circular polarization q about **k**. The interaction for other polarization states is given by a simple combination of terms like (6.11). For example, from (4.37) we see the operator for photons with linear polarization in the x-direction is just

$$H'_x(\mathbf{k},\mathbf{r}) = \frac{1}{\sqrt{2}}[H'_{-1}(\mathbf{k},\mathbf{r}) - H'_1(\mathbf{k},\mathbf{r})].$$

Of course, if the polarization state is of no interest, we use an incoherent sum of the $q = \pm 1$ terms.

The tensors (6.12) are hermitian in the sense of section (4.8). and have simple time reversal properties (section 4.9). If T_{LM} stands for any tensor of (6.12),‡

$$T^{+}_{LM} = -\boldsymbol{\theta} T_{LM}\, \boldsymbol{\theta}^{-1} = (-1)^{M-p} T_{L-M}, \tag{6.13}$$

where $p = 0$ for the magnetic and $p = 1$ for the electric terms.

For systems of more than one particle the interaction operator will be a sum of terms (6.11), one for each particle.

6.1.2. *The Matrix Elements and Transition Probabilities*

The selection rules operating for matrix elements of the tensors (6.12) are given immediately by section 4.7.2. In particular the different parity of electric and magnetic multipoles of the same rank L ensures that only one can contribute for each allowed L, to transitions between states of definite parity. For the same reason, when two multipoles of rank L and $L+1$ contribute, one will be electric and the other magnetic in order to conserve parity. The time reversal properties (6.13) ensure that reduced matrix elements of $H'_q(\mathbf{k},\mathbf{r})$ will be real (section 4.9.).

The static electric and magnetic multipole moments of a system with total angular momentum J are given by the

‡ Apparently (6.13) is not obeyed by the Q_{LM} of (6.12a), but it should be remembered that this is only an *effective* operator in the sense of equation (6.8). When the hermitian conjugate of a matrix element $\langle\beta|T|\alpha\rangle$ is taken, the factor $(E_\alpha - E_\beta)$ in (6.8) also changes sign.

diagonal, $M = J$, elements of the tensors (6.12) (summed over all particles in the system). The magnetic dipole moment is

$$\mu \equiv \beta \langle JJ|g_L L + g_s S|JJ\rangle = \sqrt{\{J/(J+1)\}}\langle J||M_1 + M_1'||J\rangle \tag{6.14}$$

while the conventional electric quadrupole moment is

$$Q = 2\langle Q_2\rangle/e,$$

$$eQ \equiv e\langle JJ|g_L(3z^2 - r^2)|JJ\rangle = 2\left(\frac{J(2J-1)}{(J+1)(2J+3)}\right)^{\frac{1}{2}} \langle J||Q_2||J\rangle. \tag{6.15}$$

The other static moments of interest are the magnetic octupole moment Ω [41] [59], and electric 2^4-pole, or hexadecapole, moment $Q^{(4)}$ [74] (this reference uses the symbol $M_{16} = 8Q^{(4)}$)

$$\Omega \equiv -\beta \langle JJ|(\tfrac{1}{2}g_L \mathbf{L} + g_s \mathbf{S}).\boldsymbol{\nabla}(r^3 C_{30})|JJ\rangle$$
$$= -\left(\frac{J(J-1)(2J-1)}{(J+1)(J+2)(2J+3)}\right)^{\frac{1}{2}} \langle J||M_3 + M_3'||J\rangle \tag{6.16}$$

$$Q^{(4)} \equiv \frac{e}{8}\langle JJ|g_L(35z^4 - 30z^2r^2 + 3r^4)|JJ\rangle$$
$$= \left(\frac{J(J-1)(2J-1)(2J-3)}{(J+2)(J+1)(2J+3)(2J+5)}\right)^{\frac{1}{2}} \langle J||Q_4||J\rangle. \tag{6.17}$$

To first order, the probability amplitude for a transition from a state $|\alpha_1 J_1 M_1\rangle$ to $|\alpha_2 J_2 M_2\rangle$ with emission of a circular polarized photon along the direction k is proportional to the matrix element of the interaction (6.11),

$$A^q_{M_1 M_2}(\mathbf{k}) = -\left(\frac{k}{2\pi\hbar}\right)^{\frac{1}{2}} \sum_{LM\pi} q^\pi \alpha^\pi_L \mathscr{D}^L_{Mq}(R)\langle J_1 M_1|T^\pi_{LM}|J_2 M_2\rangle \tag{6.18}$$

where π is e for electric ($q^\pi = 1$), π is m for magnetic ($q^\pi = q$),

$$\alpha^e_L = \frac{(ik)^L}{(2L-1)!!}\left(\frac{L+1}{2L}\right)^{\frac{1}{2}}, \quad \alpha^m_L = -i\alpha^e_L, \tag{6.19}$$

and T_{LM} stands for the tensors (6.12). If we do not observe the orientation of the spin J_2 of the final system we must sum over

M_2, and the probability $P^q_{M_1}(\mathbf{k})$ of emission from the initial substate with z—component M_1 is equal to $\sum\limits_{M_2} |A^q_{M_1M_2}(\mathbf{k})|^2$. If linear, not circular, polarization is observed, we take a coherent superposition of $q = \pm 1$ terms A^q as described in the previous section, while a polarization insensitive measurement requires an incoherent sum over q. If the radiating system is in a cylindrically symmetric environment, so that M_1 is a constant of the motion if the z-axis is chosen along the symmetry axis, the total radiative probability for photons along $\mathbf{k}$ is obtained by weighting each $P^q_{M_1}$ with the population or relative probability $\omega(M_1)$ of the substate M_1

$$P^q(\mathbf{k}) = \sum_M w(M_1)P^q_{M_1}(\mathbf{k}) = \sum_{M_1} w(M_1)\sum_{M_2} |A^q_{M_1M_2}(\mathbf{k})|^2 \tag{6.20}$$

(If the system does not possess cylindrical symmetry, J_{1z} cannot be made diagonal, and $\omega(M_1)$ has to be replaced by a more general density matrix $\rho(M_1, M_1')$: (see section (6.4).) Using the Wigner–Eckart theorem of section 4.7.1 and combining the rotation matrixes according to (2.32), we soon reduce the angular distribution (6.20) to

$$\begin{aligned} P^q(k) = \frac{k}{2\pi\hbar} \sum_{KLL'\pi\pi'} & B_K(J_1)P_K(\cos\beta)\, W(J_1J_1LL'; KJ_2) \times \\ & \times (2J_1+1)^{\frac{1}{2}} (-1)^{q+J_1-J_2+L-L'-K} q^{\pi} q^{\pi'} \langle LL'q-q|K0\rangle \times \\ & \times \langle J_1\|\alpha^{\pi}_L T^{\pi}_L\|J_2\rangle \langle J_1\|\alpha^{\pi'}_{L'} T^{\pi'}_{L'}\|J_2\rangle^*. \end{aligned} \tag{6.21}$$

β is the angle between $\mathbf{k}$ and the symmetry axis, and B_k describes the orientation of the initial system (for example, an assembly of oriented nuclei [12],

$$\begin{aligned} B_K(J) &= \sum_M w(M)(-1)^{J-M}\sqrt{(2J+1)}\langle K0|JJM-M\rangle, \\ B_0(J) &= 1. \end{aligned} \tag{6.22}$$

If the system was initially randomly oriented, so $\omega(M_1)$ is just the statistical weight $(2J_1+1)^{-1}$, only the K zero term survives, and (6.21) becomes

$$P^q(\mu) = \frac{k}{2\pi\hbar} \sum_{L\pi} |\langle J_1\|\alpha^{\pi}_L \mathbf{T}^{\pi}_L\|J_2\rangle|^2/(2L+1) \tag{6.23}$$

which is independent of the polarization q and the direction of emission $\mathbf{k}$, as expected. The total decay rate is obtained by integrating (6.21) over the direction of emission $\mathbf{k}$ and summing over the polarization q. We note that, while different multipoles interfere coherently in the angular distribution (6.21), there are no interference contributions to the total decay probability (6.23). The decay probability per second is

$$P \equiv \frac{1}{\tau} = \frac{2}{\hbar} \sum_{L\pi} \frac{k^{2L+1}(L+1)}{L[(2L-1)!!]^2} \frac{|\langle J_1 \| \mathbf{T}_L^\pi \| J_2 \rangle|^2}{(2L+1)}$$

while the 'reduced' transition probabilities introduced by Bohr and Mottelson are just

$$B_{1\to 2}(\pi L) = |\langle J_1 \| \mathbf{T}_L^\pi \| J_2 \rangle|^2 \left(\frac{2L+1}{4\pi}\right). \tag{6.24}$$

Comparison with (6.21) also shows that the matrix element which appears in formulae for the angular correlation of gamma rays is, in the conventional notation of Biedenharn and Rose [8], [19],

$$(j \| L \| j_1) = N \langle j \| \alpha_L^\pi T_L^\pi \| j_1 \rangle / (2L+1)^{\frac{1}{2}}$$

where the normalization factor N is independent of L and π.

6.1.3. *Single-particle Matrix Elements*

We can now calculate the matrix elements of the multipole tensors (6.12) between the states of a single particle in a central field. For a spin $-\frac{1}{2}$ particle with spin-orbit coupling these states are

$$|l\tfrac{1}{2}jm\rangle = \sum_\sigma |\tfrac{1}{2}\sigma\rangle u_l(r) i^l Y_{lm-\sigma}(\theta\varphi) \langle l\tfrac{1}{2}m-\sigma\sigma|jm\rangle$$

where i^l is included to give the required time reversal properties (section 4.9.). The matrix elements now have the forms discussed in Chapter V. The elements of Q_{LM} follow immediately from (5.11), while M_{LM} has the form of (5.5). The spin terms Q'_{LM} and M'_{LM} both have the form (5.12). Using these and some of the standard reduced matrix elements listed in Appendix VI we obtain

$$\langle l_1\tfrac{1}{2}j_1||\mathbf{Q}_L||l_2\tfrac{1}{2}j_2\rangle = eg_L J(L)\, b_{12}$$

$$\langle l_1\tfrac{1}{2}j_1||\mathbf{M}_L||l_2\tfrac{1}{2}j_2\rangle = 2\beta g_L i^{l_2-l_1}(-1)^{j_1-\frac{1}{2}}J(L-1)\,W(j_1l_1j_2l_2;\,\tfrac{1}{2}L)\times$$
$$\times[(2j_2+1)(2l_1+1)(2l_2+1)l_2(l_2+1)L/L+1]^{\frac{1}{2}}\begin{pmatrix} l_1 & l_2 & L \\ 0 & 1 & -1\end{pmatrix}$$

$$\langle l_1\tfrac{1}{2}j_1||\mathbf{Q}'_L||l_2\tfrac{1}{2}j_2\rangle = \tfrac{1}{2}g_s\beta kJ(L)(a_1-a_2)b_{12}/(L+1)$$

$$\langle l_1\tfrac{1}{2}j_1||\mathbf{M}'_L||l_2\tfrac{1}{2}j_2\rangle = \tfrac{1}{2}g_s\beta\, J(L-1)(L-a_1-a_2)b_{12}$$

where $b_{12} = i^{l_1-l_2}\,(-1)^{j_2-\frac{1}{2}}\,(2j_2+1)^{\frac{1}{2}}\begin{pmatrix} j_1 & j_2 & L \\ \frac{1}{2} & -\frac{1}{2} & 0\end{pmatrix},$

$$a = (l-j)(2j+1),$$

and $J(L)$ is the radial integral

$$J(L) = \int u_1(r)r^{L+2}\,u_2(r)\,dr.$$

To conserve parity $(L+l_1+l_2)$ must be even for electric operators and odd for magnetic.

For $M1$, $E1$, and $E2$ transitions the elements are very simple. For $M1$ we have $l_1 \neq l_2$ transitions are forbidden, and

$$\langle l\tfrac{1}{2}j||\mathbf{M}_1+\mathbf{M}'_1||l\tfrac{1}{2}j\rangle = [(j+1)/j]^{\frac{1}{2}}\mu(jl),$$

$$\langle l\tfrac{1}{2}j||\mathbf{M}_1+\mathbf{M}'_1||l\tfrac{1}{2}j+1\rangle = \beta(g_s-g_L)\sqrt{(l+1)/(2l+1)}.$$

$\mu(jl)$ is the Schmidt value for the magnetic moment of a particle in an orbit $(j,\,l)$ [11].

$$\mu(jl) = j\beta[g_L\pm(g_s-g_L)/(2l+1)] \quad \text{as} \quad j = l\pm\tfrac{1}{2}.$$

The dominant charge contribution to $E1$ transitions has elements

$$\langle l_1\tfrac{1}{2}j||\mathbf{Q}_1||l_2\tfrac{1}{2}j\rangle = \tfrac{1}{2}eg_L J(1)i^{l_1-l_2}/[j(j+1)]^{\frac{1}{2}},$$

$$\langle l\tfrac{1}{2}j||\mathbf{Q}_1||l+1,\,\tfrac{1}{2}j+1\rangle = -\tfrac{1}{2}ieg_L J(1)[(2j+3)/(j+1)]^{\frac{1}{2}}.$$

Similarly the charge contributions to $E2$ elements are

$$\langle l\tfrac{1}{2}j||\mathbf{Q}_2||l\tfrac{1}{2}j\rangle = -\tfrac{1}{4}eg_L J(2)\left(\frac{(2j-1)(2j+3)}{j(j+1)}\right)^{\frac{1}{2}},$$

$$\langle l_1\tfrac{1}{2}j||\mathbf{Q}_2||l_2\tfrac{1}{2}j+1\rangle = \tfrac{1}{4}eg_L J(2)i^{l_1-l_2}\left(\frac{3(2j+3)}{j(j+1)(j+2)}\right)^{\frac{1}{2}},$$

$$\langle l\tfrac{1}{2}j||\mathbf{Q}_2||l+2,\,\tfrac{1}{2},\,j+2\rangle = \tfrac{1}{4}eg_L J(2)\left(\frac{3(2j+3)(2j+5)}{2(j+1)(j+2)}\right)^{\frac{1}{2}}.$$

6.1.4. Systems of More Than One Particle

Systems of more than one particle may be treated as described in section 5.4, since the electromagnetic interaction is a one-body operator. To emphasize the possible importance of the coupling to other particles we consider here just two particles, with a transition from a state $|j_1 j J_1 M_1\rangle$ to $|j_2 j J_2 M_2\rangle$. The matrix element is given directly by (5.9). In particular let us compare the 2^L-pole transitions $(A)\, j_1 j \rightarrow j^2$ and $(B)\, j_1 j \rightarrow j_1^2$, when $J_1 = J_2 + L$. Expanding the Racah functions in (5.9) the ratio of the square of the reduced matrix for transition A to that for B is

$$\frac{(2j-J_2)!\,(J_2+2j_1+1)!}{(2j_1-J_2)!\,(J_2+2j+1)!}.$$

The interesting cases arise when j_1 and J_1 are large, and j, J_2 are small. For example take $j_1 = \frac{13}{2}$, $j = \frac{3}{2}$, $J_1 = 7$, and $J_2 = 2$, so that $L = 5$. Then, although the same single particle matrix element $\langle\frac{3}{2}\|T_5\|\frac{13}{2}\rangle$ is involved in both transitions, the ratio of their probabilities is 1/728, while transition A is reduced by the angular momentum coupling to 1/2070 the intensity of the single particle $L = 5$ transition $\frac{3}{2} \rightarrow \frac{13}{2}$.

More detailed considerations of transitions between many-particle states have been given elsewhere for both $j-j$ and $L-S$ coupling [14], [43], [46], [49]. Some interesting results arise from the general properties of matrix elements discussed in Chapter V. For example, the relative probabilities (or 'line strengths' [17]) for electric transitions between the components of two Russell-Saunders multiplets follow immediately from the decoupling relation (5.9). The reduced transition probability $B(EL)$ of (6.24) between the $L-S$ states $|\alpha_1 L_1 S_1 J_1\rangle$ and $|\alpha_2 L_2 S_2 J_2\rangle$ becomes

$$4\pi B_{1\rightarrow 2}(EL) = \delta_{S_1 S_2}(2L_1+1)(2J_2+1)(2L+1)\times$$
$$\times W^2(J_1 J_2 L_1 L_2;\, LS_1)\big|\langle\alpha_1 L_1\|\mathbf{T}_L^e\|\alpha_2 L_2\rangle\big|. \quad (6.25)$$

So between states of the same two multiplets, but with different J_1 and J_2, we have the purely geometrical ratio

$$\frac{B_{J_1'\rightarrow J_2'}(EL)}{B_{J_1\rightarrow J_2}(EL)} = \frac{(2J_2'+1)\,W^2(J_1' J_2' L_1 L_2;\, LS_1)}{(2J_2+1)\,W^2(J_1 J_2 L_1 L_2;\, LS_1)}.$$

Again, if we sum (6.25) over final angular momenta J_2, we get

$$4\pi\sum_{J_2} B_{J_1\to J_2}(EL) = (2L+1)|\langle\alpha_1 L_1\|\mathbf{T}^e_L\|\alpha_2 L_2\rangle|^2.$$

So, between two $L-S$ multiplets the total intensity from an initial Zeeman level with J_1, M_1 is independent of both J_1 and M_1. Hence the *sum* of intensities from the Zeeman components of a given J_1 is proportional to its statistical weight $(2J_1+1)$. Conversely, the sum of intensities feeding the Zeeman levels of a given final J_2 is proportional to $(2J_2+1)$.

6.2. Interaction between two Systems

In this and the following section we shall discuss the interaction of two systems, each of which may itself possess some structure. If the combined system is isolated, the Hamiltonian is invariant under rotations and any interaction terms coupling parts 1 and 2 must be expressible as scalar products of tensors as in (4.4),

$$V(1, 2) = \sum_K \mathbf{R}_K(1)\,.\,\mathbf{S}_K(2). \tag{6.26}$$

The matrix elements of such products are given by (5.13) in terms of the matrix elements for the component parts 1 and 2; it merely remains to discuss some specific forms for the tensors $\mathbf{R}_K$ and $\mathbf{S}_K$.‡ We shall also mention briefly a case, anisotropic hyperfine structure, where the effective interaction has to be represented by a tensor product rather than the scalar form (6.26).

6.2.1. Interaction of Nuclei with Atomic Fields

The interaction between two charge distributions was discussed in section 4.10.5, and the form (6.26) found for both

‡ With a second quantization treatment of radiation fields, as mentioned in section 4.10.4, radiative transitions are also induced by scalar product interaction terms like (6.26). For example, the electromagnetic interaction will have the same form (6.11) as in the previous section, except that, following (4.50), each multipole tensor will be associated with a photon creation or annihilation operator in the proper scalar combination. Matrix elements are then taken between product states for field and radiating system.

the electric interaction (4.59) and the magnetic interaction (4.66). Important dynamic applications of these interactions occur in the theory of internal conversion [53], [55], whereby a nucleus transfers excitation energy to an atomic electron, and Coulomb excitation of a nucleus by bombardment with charged particles [1]. Here we shall consider briefly the stationery effect of the interaction between a nucleus and its electronic environment which leads to the hyperfine splitting of atomic spectral lines [41] [59]. Ignoring penetration into the nucleus by the electrons (so $r_e > r_n$), the interaction for isolated atoms is given by (4.59) and (4.67)

$$V(e, n) = \sum_{LM} [Q^*_{LM}(e) Q_{LM}(n) + M^*_{LM}(e) M_{LM}(n)]. \quad (6.27)$$

The nuclear electric moments $Q(n)$ are just those defined by (6.12a) in the previous section, while the electric moments $M(n)$ are the sum of the charge and spin contributions (6.12b) and (6.12d). (Summed, of course, over all nucleons.) The corresponding moments $Q(e)$ and $M(e)$ for the electron cloud are given by (4.60) and (4.68), and are similar to the nuclear moments except in radial dependence. The quadrupole operator $Q_{20}(e)$, for example, is just $\frac{1}{2}\partial^2 V/\partial z^2$ at the origin, where V is the electrostatic potential due to the electrons.

The small contributions from $r_e < r_n$ may be deduced from section 4.10.5, for example the higher terms in the expansion (4.58).

The first-order hyperfine energy shift W_{IJF} for an atomic level with total angular momentum F (vector sum of nuclear I and electronic J) is just the expectation value of $V(e, n)$‡. It

‡ Second order effects of low multipoles may be important in the interpretation of higher multipoles; for example, the second order term for the dipole $L = 1$ has a part which looks like a first order quadrupole, $L = 2$, term [41], [59]. Also the much closer approach of the meson in μ-mesic atoms allows higher order effects to be important which involve excitation of low-lying nuclear levels (nuclear 'polarization') through the off-diagonal elements of $V(e, n)$ [79].

follows immediately from (5.13) that

$$W_{IJF} = \langle IJF|V(e,n)|IJF\rangle$$
$$= \sum_{L\pi} (-1)^{I+J-F} [(2I+1)(2J+1)]^{\frac{1}{2}} W(IJIJ;\, FL) \times$$
$$\times \langle I\|\mathbf{T}_L^{\pi}(n)\|I\rangle\langle J\|\mathbf{T}_L^{\pi}(e)\|J\rangle \qquad (6.28)$$

where $\mathbf{T}^e = \mathbf{Q}$, $\mathbf{T}^m = \mathbf{M}$, and parity conservation ensures that only $\mathbf{Q}_L$ with L even, and $\mathbf{M}_L$ with L odd, can contribute. We can invert (6.28) to express the reduced matrix elements in terms of the observed W_{IJF}, using the orthogonality (3.17) of the Racah coefficients

$$[(2I+1)(2J+1)]^{\frac{1}{2}}\langle I\|T_L^{\pi}\|I\rangle\langle J\|T_L^{\pi}\|J\rangle$$
$$= \sum_F (-1)^{I+J-F}[(2F+1)(2L+1)]\; W(IJIJ,\, FL)\, W_{IJF}. \qquad (6.29)$$

The evaluation of the reduced matrix elements has been discussed in the previous section and elsewhere [11], [41], [59].

Allied problems, such as in molecular spectra, or with the application of external magnetic fields, can be treated by similar techniques [41], [50]. We shall confine ourselves to a few remarks on an ion which is not isolated, but situated in a crystalline electric field. Such an electrostatic field can be expanded in solid harmonics about the position of the ion (following section 4.10)

$$V = \sum_{kq} a_{kq} r^k C_{kq}(\theta, \varphi). \qquad (6.29)$$

The coefficients a_{kq} are determined by the nature and symmetry of the crystal lattice [10]. The matrix elements of (6.29) are readily evaluated using the Wigner–Eckart theorem and the techniques of Chapter V [25]. In the 'single-electron' approximation for the ionic states this is almost trivial. An alternative method for diagonal matrix elements between many-electron states is to find an 'operator equivalent' for each term in (6.29) [68]. Between states of the same J we use J_x, J_y, and J_z to construct operators which have the same matrix elements as (6.29); for example

$$r^2C_{20} \rightarrow \alpha_2\langle r^2\rangle[3J_z^2 - J(J+1)].$$

Then $\alpha_2\langle r^2\rangle$ is proportional to the reduced matrix element of r^2C_2, and techniques have been devised for their evaluation [68]. The other part of the operator equivalent gives matrix elements proportional to the Clebsch–Gordan coefficient of the Wigner–Eckart theorem; for the example above,

$$\langle JM|3J_z^2-J(J+1)|JM'\rangle$$
$$= \delta(MM')[J(J+1)(2J-1)(2J+3)]^{\frac{1}{2}}\langle JM|J2M0\rangle.$$

This approach, however, will not give off-diagonal matrix elements.

This overall crystal field polarizes individual ions and induces anisotropic hyperfine structure. The environment is no longer spherically symmetric and the effective nucleus-electron interaction can include tensor products of higher rank than the scalars in (6.27). A particular example is the spin-Hamiltonian of Abragam and Pryce [10] which includes a term

$$H_{SI} = AS_z I_z+BS_x I_x+CS_y I_y$$
$$= AS_0I_0+\tfrac{1}{2}(B-C)(S_1I_1+S_{-1}I_{-1})-\tfrac{1}{2}(B+C)(S_1I_{-1}+S_{-1}I_1).$$

$\mathbf{S}$ is an effective spin of the atomic electrons. The symmetry characteristics of H_{SI} are made evident if we express it in terms of product tensors (4.6) built from $\mathbf{S}$ and $\mathbf{I}$.

$$H_{SI} = \alpha\mathbf{S}\,.\,\mathbf{I}+\beta T_{20}(\mathbf{S},\,\mathbf{I})+\gamma[T_{22}(\mathbf{S},\,\mathbf{I})+T_{2\,-2}(\mathbf{S},\,\mathbf{I})] \tag{6.30}$$

where

$$\alpha = \tfrac{1}{3}(A+B+C), \qquad \beta = \frac{1}{\sqrt{6}}(2A-B-C), \qquad \gamma = \tfrac{1}{2}(B-C).$$

The effect of the anisotropy is to introduce second rank tensors into the interaction which will connect states of different F; that is, F is no longer a good quantum number in the presence of this interaction. However, if axial symmetry is restored by putting $B = C$, the $T_{2\pm2}$ terms vanish and $F_z = I_z+S_z$ does remain a constant of the motion. H_{SI} is, of course, diagonal in S and I even though it mixes different S_z, I_z values.

This interaction is the basis of the Bleaney method of orienting atomic nuclei [12]. The crystal field orients the ionic electron clouds which in turn, through the anisotropic interaction H_{SI}, orient the nuclei. Of course, to be effective the sample must be cooled to a temperature T such that $kT \sim$ the hyperfine splittings induced by H_{SI}. Because H_{SI} is symmetric in $\pm I_z$, only nuclear 'alignment' can be produced; that is, only even-order polarization moments are induced, and $\langle I_z^n \rangle$ vanishes for n odd.

6.3. Interactions between two Particles in a Central Field

A particular case of the interaction of two systems which is sufficiently important in the theory of atomic and nuclear structure to warrant separate treatment, is the mutual-interaction of two particles moving in orbits in a central field. The particles may possess spin. Then the various interaction terms, although scalar in the overall product space of spins and orbits, may be classified according to their properties under spatial rotations in the spin and orbit spaces separately. If the position coordinates are denoted $\mathbf{r}_1$ and $\mathbf{r}_2$ and the spin operators $\mathbf{s}_1$ and $\mathbf{s}_2$, the interaction can be written in the scalar product form

$$V(12) = \sum_K \mathbf{R}_K(\mathbf{r}_1, \mathbf{r}_2) \,.\, \mathbf{S}_K(\mathbf{s}_1, \mathbf{s}_2). \tag{6.31}$$

$\mathbf{R}_K(r_1 r_2)$ operates only on the position coordinates, $\mathbf{S}_K(s_1, s_2)$ only on the spins; each is built up from its arguments in the way discussed in Chapter IV.

A nuclear system of neutrons and protons may be described by the isotopic spin formalism; the extension of (6.31) and the discussion below to this case is straightforward (see section 5.4.2 and [26]) and will not be considered here.

6.3.1. *Spinless Particles and Central Forces*

Simplest is the scalar $K = 0$ (or central) force acting between particles without spins. This will be just a function of the distance r_{12} between the particles, $J(r_{12})$ say, where

$\mathbf{r}_{12} = \mathbf{r}_1 - \mathbf{r}_2$. The matrix element of this may be evaluated by the Slater method [26], [69] by expressing $J(r_{12})$ itself in the form (6.26). We write

$$J(r_{12}) = \sum_k J_k(r_1, r_2) P_k(\cos\omega) \tag{6.32}$$

where

$$J_k(r_1, r_2) = \tfrac{1}{2}(2k+1)\int_0^\pi J(r_{12}) P_k(\cos\omega) \sin\omega \, d\omega. \tag{6.33}$$

Expressions for J_k for various interactions have been given [69], and the particular case $J = e^2/r_{12}$ has been discussed in Chapter IV. ω is the angle between $\mathbf{r}_1$ and $\mathbf{r}_2$. The Legendre polynomials may be expanded by the addition theorem (2.25), $P_k(\cos\omega) = \mathbf{C}_k(1) \cdot \mathbf{C}_k(2)$. If the particles are spinless and their orbital angular momenta are coupled to a resultant L, we may immediately carry out the angular integrations of the matrix element of (6.32) using (5.14). In the usual notation, and using single particle wave functions

$$|lm\rangle = u_{nl}(r) Y_{lm}(\theta, \varphi)$$

$$\langle l_1 l_2 LM | J(r_{12}) | l_1' l_2' L' M' \rangle$$
$$= \delta(LL')\delta(MM') \sum_k f_k(l_1 l_2 l_1' l_2'; L) R^{(k)}(l_1 l_2 l_1' l_2') \tag{6.34}$$

where f_k is just $\langle l_1 l_2 L | C_k(1) . C_k(2) | l_1' l_2' L \rangle$ given in equation (5.14). The f_k have the following symmetries:

$$f_k(l_1 l_2 l_1' l_2'; L) = f_k(l_1' l_2' l_1 l_2; L) = f_k(l_2 l_1 l_2' l_1'; L)$$

and (6.35)

$$f_0 = \delta(l_1 l_1')\, \delta(l_2 l_2').$$

The condition that both $(k+l_1+l_1')$ and $(k+l_2+l_2')$ be even ensures that $J(r_{12})$ does not connect states of different total parity, and k is limited to $\leqslant l_1+l_1'$ and $\leqslant l_2+l_2'$. The R^k are Slater radial integrals, parameters independent of the total angular momentum value L,

$$R^k(l_1 l_2 l_1' l_2') = \int r_1^2 \, dr_1 \int r_2^2 \, dr_2 \, J_k(r_1, r_2) u_{l_1}(r_1) u_{l'_1}(r_1) u_{l_2}(r_2) u_{l'_2}(r_2). \tag{6.36}$$

$u_{nl}(r)$ is the single particle radial wave function; the indices n denoting the principal quantum numbers of the orbits have been suppressed for clarity. In the limit of very long range (compared to the extension of the wave functions $u(r)$) $J(r_{12})$ becomes constant, independent of $\mathbf{r}_1$ and $\mathbf{r}_2$, and (6.36) ensures that J_k vanishes unless $k = 0$. In this limit all matrix elements vanish which are off-diagonal in the single particle orbits, and the others are independent of L:

$$\langle l_1 l_2 LM|J(r_{12})|l_1' l_2' LM\rangle \to \delta(l_1 l_1')\,\delta(l_2 l_2')\,R^0(l_1 l_2 l_1 l_2). \quad (6.37)$$

The other limit of zero-range is of interest for nuclear structure. The expansion (6.32) for a delta function is

$$\delta(\mathbf{r}_1-\mathbf{r}_2) = \frac{1}{2\pi r_1^2}\,\delta(r_1-r_2)\,\delta(\cos\omega-1)$$

$$= \frac{1}{4\pi r_1^2}\,\delta(r_1-r_2)\sum_k (2k+1)\mathbf{C}_k(1)\,.\,\mathbf{C}_k(2), \quad (6.38)$$

then immediately in this limit

$$J_k(r_1, r_2) = A\frac{2k+1}{4\pi r_1^2}\,\delta(r_1-r_2) = (2k+1)J_0(r_1, r_2), \quad (6.39)$$

if $J(r_{12}) = A\,\delta(\mathbf{r}_1-\mathbf{r}_2)$. Thus we have $R^{(k)} = (2k+1)R^{(0)}$ and this allows us to carry out the sum over k in (6.34) using (3.11). The zero range matrix element becomes

$$\langle l_1 l_2 LM|J(r_{12})|l_1' l_2' LM\rangle = P(l_1 l_2 l_1' l_2';\, L)R^{(0)} \quad (6.40)$$

where

$$P(l_1 l_2 l_1' l_2';\, L)$$
$$= \begin{pmatrix} l_1 & l_2 & L \\ 0 & 0 & 0 \end{pmatrix}\begin{pmatrix} l_1' & l_2' & L \\ 0 & 0 & 0 \end{pmatrix}[(2l_1+1)(2l_2+1)(2l_1'+1)(2l_2'+1)]^{\frac{1}{2}}$$

which, of course, vanishes unless (l_1+l_2+L) and $(l_1'+l_2'+L)$ are both even. The radial overlap $R^{(0)}$ has become simply

$$R^{(0)} = \frac{A}{4\pi}\int r^2\,dr\,u_{l_1}(r)u_{l_2}(r)u_{l_1'}(r)u_{l_2'}(r). \quad (6.41)$$

Occasionally it is more convenient to work with the uncoupled matrix elements, $\langle l_1 m_1 l_2 m_2 | J | l_1' m_1' l_2' m_2' \rangle$. This will be so if the environment is not spherically symmetric, such as for the motion of individual nucleons in a strongly deformed nucleus [13], so the total L is not a constant of the motion any more than the individual l. The same multipole expansion (6.34) is used, and the angle integrations carried out by use of (4.16)

$$\langle l_1 m_1 l_2 m_2 | J(r_{12}) | l_1' m_1' l_2' m_2' \rangle = \delta(m_1+m_2,\, m_1'+m_2')\times \times \sum_k c^k(l_1 m_1 l_1' m_1')\, c^k(l_2 m_2 l_2' m_2') R^{(k)}; \qquad (6.42)$$

$$c^k(lml'm') = [(2l+1)(2l'+1)]^{\frac{1}{2}}(-1)^{m'}\begin{pmatrix} l & l' & k \\ 0 & 0 & 0 \end{pmatrix}\begin{pmatrix} l & l' & k \\ m & -m' & m'-m \end{pmatrix} = (-1)^{m-m'}\, c^k(l'm'lm).$$

Values of c^k have been tabulated [17], [64].

Finally, it should be remarked that a technique different from the application of (6.32) has been developed by Talmi [32], [70], [73] for use when the functions $u_{nl}(r)$ are eigenfunctions of an harmonic oscillator potential well.

6.3.2. *Particles with Spin; Central Forces*

When the particles have spin, $K \neq 0$ terms may appear in (6.31), and the scalar $K = 0$ matrix element is itself modified. Because of the form of (6.31) it is most convenient to use an L–S or Russel-Saunders coupling scheme, when we get an example of the general matrix element (5.13)‡.

$$\langle LSJM | V(12) | L'S'J'M' \rangle = \delta(JJ')\, \delta(MM')(-1)^{J-L-S'} \times \times \sum_K W(LL'SS';KJ)\sqrt{(2L+1)}\langle L \| R_K(r_1, r_2) \| L' \rangle \times \times \sqrt{(2S+1)}\langle S \| S_K(s_1, s_2) \| S' \rangle. \qquad (6.43)$$

If the scalar part ($K = 0$) of $V(1,2)$ is spin independent

‡ Matrix elements in the $j-j$ coupling representation are easily obtained from (6.43) by using the transformation (3.23), (3.24)

$$\langle j_1 j_2 J | V | j_1' j_2' J \rangle = \sum_{LL'SS'} \langle LSJ | V | L'S'J \rangle \times \times \langle (l_1 s_1) j_1, (l_2 s_2) j_2; J | (l_1 l_2) L, (s_1 s_2) S; J \rangle \langle (l_1' s_1) j_1', (l_2' s_2) j_2'; J | (l_1' l_2') L, (s_1 s_2) S'; J \rangle.$$

($\mathbf{S}_0(\mathbf{s}_1\mathbf{s}_2) = 1$, $\langle S\|1\|S\rangle = 1$), the decoupling factor represented by the Racah function gives unity. In addition, however, there may be a scalar spin interaction,

$$\mathbf{S}_0 = \mathbf{s}_1 \cdot \mathbf{s}_2,$$

$$\langle S\|\mathbf{s}_1 \cdot \mathbf{s}_2\|S\rangle = \tfrac{1}{2}[S(S+1)-s_1(s_1+1)-s_2(s_2+1)]. \tag{6.44}$$

When $s_1 = s_2 = s$ we may introduce also the spin exchange operator P^s. Because of the symmetry induced by the Clebsch–Gordan coefficient in the coupled state $|ssSM\rangle$ when the angular momenta are equal (see section 2.7.3.), P^s has the eigenvalues $(-)^{S-2s}$,

$$P^s|ssSM\rangle = (-1)^{S-2s}|ssSM\rangle \tag{6.45}$$

hence

$$\langle S\|P^s\|S\rangle = (-1)^{S-2s}.$$

In particular, when $s = \frac{1}{2}$ we can use (6.45) to give a representation for P^s:

$$P^s = \tfrac{1}{2}+2\mathbf{s}_1 \cdot \mathbf{s}_2 = \tfrac{1}{2}(1+\boldsymbol{\sigma}_1 \cdot \boldsymbol{\sigma}_2). \tag{6.46}$$

The second form in terms of the Pauli spin matrices follows since $\mathbf{s} = \frac{1}{2}\boldsymbol{\sigma}$ here. This allows us to evaluate the expectation value of the spin exchange operator for n spin-$\frac{1}{2}$ particles, $M = \frac{1}{2}\sum_{ij} P^s_{ij}$, (known as Hund's operator) without recourse to the methods of Section 5.3. Because of the meaning of P^s in (6.45), the value of M is the difference between the numbers of symmetric and antisymmetric pairs of spins. Using 6.46 and remembering that

$$\mathbf{S}^2 = (\sum_i \mathbf{s}_i)^2 = \sum_i \mathbf{s}_i^2+\sum_{i\neq j} \mathbf{s}_i \cdot \mathbf{s}_j$$

we get

$$M = \tfrac{1}{2}[\tfrac{1}{2}n(n-1)+2\mathbf{S}^2-2\sum_i \mathbf{s}_i^2]$$

and since $s = \frac{1}{2}$, for a state of total spin S

$$\langle S|M|S\rangle = \tfrac{1}{2}[\tfrac{1}{2}n(n-4)+2S(S+1)]. \tag{6.47}$$

6.3.3. Vector and Tensor Forces

The terms with $K = 1$ and 2 in (6.31) are referred to as vector and tensor forces, respectively. The only vector term linear in the spins and momenta which satisfies the requirements of symmetry and space- and time-inversion invariance is (Wigner [76])

$$J(r_{12})\mathbf{L}(1,2)\,.\,\mathbf{S}, \tag{6.48}$$

where

$$\mathbf{S} = \mathbf{s}_1+\mathbf{s}_2$$

and

$$\mathbf{L}(1,2) = \tfrac{1}{2}(\mathbf{r}_1-\mathbf{r}_2)\wedge(\mathbf{p}_1-\mathbf{p}_2) \tag{6.49}$$

is the *relative* orbital angular momentum. The matrix elements of $\mathbf{S}_1(s_1 s_2) \equiv \mathbf{S}$ are simply dealt with, but $\mathbf{R}(r_1 r_2) = J(r_{12})$ $\mathbf{L}(1, 2)$ requires more attention. The terms in the expansion (6.32) of $J(r_{12})$ have to be coupled to those from the corresponding expansion of $\mathbf{L}(1, 2)$,

$$\mathbf{L}(1,2) = \tfrac{1}{2}\{\mathbf{L}(1)+\mathbf{L}(2)-\mathbf{r}_1\wedge\mathbf{p}_2+\mathbf{p}_1\wedge\mathbf{r}_2\}. \tag{6.50}$$

The first two terms, the one-particle orbital operators, are straightforward. To manipulate the other two terms we use the relations $\mathbf{p} = -i\boldsymbol{\nabla}$ and

$$\boldsymbol{\nabla} = \mathbf{C}_1\frac{\partial}{\partial r}-\frac{i}{r}\mathbf{C}_1\wedge\mathbf{L}, \tag{6.51}$$

where $\mathbf{C}_1$ is the unit vector along $\mathbf{r}$; $\mathbf{L}(1, 2)$ then becomes

$$\begin{aligned}\mathbf{L}(1,2) = \tfrac{1}{2}\Big\{&\mathbf{L}(1)+\mathbf{L}(2)+i\mathbf{C}_1(1)\wedge\mathbf{C}_1(2)\Big(r_1\frac{\partial}{\partial r_2}-r_2\frac{\partial}{\partial r_1}\Big)\\ &+\mathbf{C}_1(1)\wedge(\mathbf{C}_1(2)\wedge\mathbf{L}(2))\frac{r_1}{r_2}\\ &+\mathbf{C}_1(2)\wedge(\mathbf{C}_1(1)\wedge\mathbf{L}(1))\frac{r_2}{r_1}\Big\}.\end{aligned} \tag{6.52}$$

When combined with the expansion (6.31) of $J(r_{12})$, this is a series of tensor products. These may be evaluated by straightforward, although somewhat lengthy, applications of the techniques described here and in the previous chapter [23], [37].

The tensor force with $K = 2$ is simpler because of the

absence of the gradient operator. It is usually written

$$J(r_{12})S(12)$$

where

$$\begin{aligned} S(12) &= (\mathbf{s}_1 \cdot \boldsymbol{\rho})(\mathbf{s}_2 \cdot \boldsymbol{\rho}) - \tfrac{1}{3}(\mathbf{s}_1 \cdot \mathbf{s}_2) \\ &= \mathbf{R}_2(\boldsymbol{\rho}) \cdot \mathbf{S}_2(\mathbf{s}_1\mathbf{s}_2), \end{aligned} \tag{6.53}$$

$\boldsymbol{\rho}$ being the unit vector along $\mathbf{r}_{12}$. The interaction of two magnetic dipoles $\boldsymbol{\mu}_1$ and $\boldsymbol{\mu}_2$, for example, would have $J(r_{12}) = g_1 g_2/r_{12}^3$, where g_i is the g-factor for the ith particle, $\boldsymbol{\mu}_i = g_i \mathbf{s}_i$. From (5.12) the matrix elements of the spin tensor for two spin $\frac{1}{2}$ particles are‡

$$\langle S \| \mathbf{S}_2(\mathbf{s}_1\mathbf{s}_2) \| S' \rangle = \delta(SS')\delta(S1)\sqrt{\tfrac{5}{12}}. \tag{6.54}$$

It is diagonal and vanishes for singlet, $S = 0$, states. The coordinate tensor is just a spherical harmonic and is readily re-written in terms of tensors acting on the coordinates $\mathbf{r}_1$ and $\mathbf{r}_2$ separately,

$$r_{12}^2 \mathbf{C}_2(\boldsymbol{\rho}) = r_1^2 \mathbf{C}_2(1) + r_2^2 \mathbf{C}_2(2) - \sqrt{6}\,\mathbf{R}_2(\mathbf{r}_1\mathbf{r}_2). \tag{6.55}$$

It is convenient to expand, not $J(r_{12})$, but

$$\frac{1}{r_{12}^2} J_{12} = \sum_k I_k(r_1 r_2)\mathbf{C}_k(1) \cdot \mathbf{C}_k(2). \tag{6.56}$$

Again we have a series of tensor products to be evaluated by repeated application of the standard techniques. Because of the $1/r_{12}^2$ in (6.56), the separate radial integrals contain divergent parts, but these cancel in the final result [23], [36].

Again, when the radial functions for the single particle motion are eigen functions of an harmonic oscillator well, the technique of Talmi [71] greatly simplifies the matrix elements for vector and tensor forces.

6.4. Multipole Expansion of the Density Matrix

It is beyond the scope of this book to give a detailed discussion of either the basic theory or the widespread

‡ $S(12)$ is defined in (6.53) with the actual spin operators $\mathbf{s}$. Thus for spin $\frac{1}{2}$ it is $\frac{1}{4}$ of the tensor operator often defined using instead the Pauli spin operators, $\boldsymbol{\sigma} = 2\mathbf{s}$.

applications of the density or statistical matrix and its associated statistical tensors. We can only outline the techniques, briefly referring the reader to the literature for more details [8], [19], [30], [31], [72].

6.4.1. The Density Matrix

If all the (identical) component systems of an assembly are described by the same wave function

$$|\rangle = \sum_n a_n|n\rangle, \qquad \sum_n |a_n|^2 = 1 \tag{6.57}$$

the assembly is said to be in a *pure* state, and the density matrix $\boldsymbol{\rho}$ for the assembly is defined by

$$\rho_{nm} = a_n a_m^*, \qquad tr\boldsymbol{\rho} = 1. \tag{6.58}$$

It often happens that we do not have such complete information about the assembly. An example is our knowledge of the individual spin orientations in a partially polarized beam of particles or assembly of non-interacting atoms or nuclei. This more general *mixed* state can always be described as a weighted mixture of the pure states (6.57) and (6.58), so that $\boldsymbol{\rho}$ becomes an average over all N component systems of the assembly

$$\rho_{nm} = \frac{1}{N}\sum_{\nu=1}^{N} a_n(\nu)a_m(\nu)^* \equiv \overline{a_n a_m^*}. \tag{6.59}$$

Clearly $\boldsymbol{\rho}$ is Hermitian, and from (6.57) and (6.59) the expectation value in the assembly of some operator $\mathbf{O}$ is given by

$$\langle\mathbf{O}\rangle = \sum_{nm} \boldsymbol{\rho}_{nm} O_{mn} \equiv tr(\boldsymbol{\rho}\mathbf{O}). \tag{6.60}$$

If the representation we have chosen makes $\boldsymbol{\rho}$ diagonal (for example if $\boldsymbol{\rho}$ describes a paramagnetic gas in a uniform magnetic field, and m is the component of spin along the field direction) we can write

$$\rho_{nm} = w(m)\delta_{mn}. \tag{6.61}$$

$w(m)$ is then the population function, or probability of

finding one of the component systems in the state $|m\rangle$. (6.60) then takes a particularly transparent form

$$\langle \mathbf{O} \rangle = \sum_m w(m)\, O_{mm}. \tag{6.62}$$

6.4.2. *Statistical Tensors or State Multipoles*

When the angular symmetries of the assembly are of interest (that is, its properties under spatial rotations), the natural choice for the basis $|n\rangle$ in (6.57) are the eigenstates $|\alpha JM\rangle$ of the angular momentum J and its z-component M (α denoting any other quantum numbers required). From (6.58) and (6.59) we see $\boldsymbol{\rho}$ transforms like an operator under a change of representation $\boldsymbol{\rho}' = R^{+}\boldsymbol{\rho}R$ if $|n\rangle' = R|n\rangle$. In particular, upon rotating the coordinate axes through Euler angles $(\alpha\beta\gamma)$, an element of $\boldsymbol{\rho}$ referred to the new axes is expressed in terms of those referred to the old by

$$\rho'_{\alpha JM,\alpha' J'M'} = \sum_{\mu\mu'} (\mathscr{D}^{J}_{\mu M}(\alpha\beta\gamma))^{*} \rho_{\alpha J\mu,\alpha' J'\mu'} \mathscr{D}^{J'}_{\mu' M'}(\alpha\beta\gamma). \tag{6.63}$$

It is more convenient however to choose linear combinations of the elements (6.63) that form an irreducible representation; that is, carry out a multipole expansion of the elements of $\boldsymbol{\rho}$. This is clearly just a matter of vector addition since from (6.59) the elements of $\boldsymbol{\rho}$ are bilinear combinations of tensor components $a_{\alpha JM}$. So we can write

$$\rho_{\alpha JM,\alpha' J'M'} = \sum_{KQ} \rho_{KQ}(\alpha J, \alpha' J')\langle KQ|JJ', -MM'\rangle(-1)^{K-J'-M}. \tag{6.64}$$

The ρ_{KQ} were called *statistical tensors* [28], or *state multipoles* [29], by Fano,‡ and by inverting (6.64) may be written

$$\rho_{KQ}(\alpha J, \alpha' J') = \sum_{M} \rho_{\alpha JM,\alpha' J'M'}(-1)^{K-J'-M}\langle JJ' -MM'|KQ\rangle \tag{6.65}$$

‡ Unfortunately a variety of notations and definitions are in use for the statistical tensors. Ours agrees with some others [8], [6], [28], [29]. Several authors define them so that the tensor of rank K behaves under rotations like Y^{*}_{K}, contragrediently to ours [31], [19]. This is appropriate if the statistical tensors are regarded as the *coefficients* in an expansion of the density operator into a set of multipole operators.

where

$$\rho_{00}(\alpha J,\alpha' J') = \delta_{JJ'}(2J+1)^{-\frac{1}{2}} \sum_{M} \rho_{\alpha JM,\alpha' JM}.$$

When α, J have unique values, the latter becomes

$$\rho_{00} = (2J+1)^{-\frac{1}{2}} tr \boldsymbol{\rho}.$$

From its construction ρ_{KQ} is a component of a spherical tensor,

$$\rho'_{KQ'} = \sum_{Q} \rho_{KQ} \mathscr{D}^{K}_{QQ'}(\alpha\beta\gamma) \tag{6.66}$$

which may be confirmed directly from (6.63) using (2.31). The Hermitian property of ρ reappears as

$$\rho_{KQ}(\alpha J, \alpha' J') = (-1)^{J'-J+Q} \rho_{K-Q}(\alpha' J', \alpha J)^{*} \tag{6.67}$$

The definition (6.65) shows we can define tensors of rank K where $|J-J'| \leqslant K \leqslant J+J'$. It is then consistent to talk of the assembly possessing dipolarization ($K = 1$), quadripolarization ($K = 2$), ... 2^K-polarization, if the corresponding tensors do not vanish. Only for spin-$\frac{1}{2}$ systems does the term 'polarization' have a unique meaning.‡ The monopole tensor ($K = 0$) is merely a normalization constant. For systems of sharp J the other low rank tensors have simple interpretations [6]. For example the dipole tensor is just the expectation value of the angular momentum operator

$$\rho_{1Q}(JJ) = \langle J_Q \rangle \, [\tfrac{1}{3}J(J+1)(2J+1)]^{-\frac{1}{2}},$$

and the $Q = 0$ component of the quadrupole tensor is

$$\rho_{20}(JJ) = \langle 3J_z^2 - J(J+1) \rangle [\tfrac{1}{5}J(J+1)(2J-1)(2J+1)(2J+3)]^{-\frac{1}{2}}.$$

If it is possible by a rotation of axes (6.66) to make $\boldsymbol{\rho}$ diagonal in M, then with this choice of axes only tensors ρ_{KQ} with $Q = 0$ do not vanish. Physically this implies an axis of cylindrical symmetry for the assembly (for example, the direction of an applied magnetic field in a gas of paramagnetic atoms). If there is no such axis there will always be tensors with $Q \neq 0$. If the assembly is isotropic, so that $\boldsymbol{\rho}$ does not depend on M or M', (6.65) and the orthogonality of the Clebsch–Gordan

‡ Some authors restrict the use of 'polarization' to systems with odd-order polarization, using 'alignment' for even-order. Dipolarization is also called 'vector polarization.'

coefficients shows that only the monopole ($K = 0$) tensor does not vanish, and the assembly is said to be unpolarised.

If some operator **O** is expressed as a sum of spherical tensor operators

$$\mathbf{O} = \sum_{KQ} a_{KQ} O_{KQ},$$

its expectation value in the assembly is found to be

$$\begin{aligned} \langle \mathbf{O} \rangle &= \mathrm{tr}(\rho \mathbf{O}) \\ &= \sum \left(\frac{2J'+1}{2K+1}\right)^{\frac{1}{2}} a_{KQ} \rho_{KQ}(\alpha J, \alpha J') \langle \alpha' J' \| \mathbf{O}_K \| \alpha J \rangle. \end{aligned} \tag{6.68}$$

The sum runs over $K, Q, \alpha, \alpha', J, J'$. We see that all dependence on the orientation of the coordinate axes (that is, on Q) is thrown into the statistical tensors ρ_{KQ}. It is this which makes these tensors such a convenient way of describing the assembly. Further, we see from (6.68) that the expectation value of a multipole operator $\mathbf{O}_K$ of rank K depends only on the statistical tensor of the same rank.

When the systems of the assembly are composite, so that the angular momentum **J** is the resultant of two or more component angular momenta (for example the spin and orbital momenta of a particle), the statistical tensors will show a corresponding structure. They will be expressible in terms of the tensors for the component parts, coupled in a way closely analogous to the matrix elements discussed in section 5.3 [6], [19].

6.4.3. Development in Time

The statistical tensors are a very convenient way of expressing the 'angular information' contained in the assembly. Now we look at the changes that can occur as the system evolves from time t_0 to t_1, such as in a nuclear reaction or radioactive decay. These are described by the unitary transformation induced by the time-development operator of Dirac [20],

$$|t_1\rangle = U(t_1 t_0)|t_0\rangle$$

so that

$$\boldsymbol{\rho}(t_1) = U(t_1 t_0)\boldsymbol{\rho}(t_0)U(t_1 t_0)^{+}. \tag{6.69}$$

As a special case we have the S-matrix [33] or collision matrix [44], $S = U(\infty, -\infty)$, connecting the initial and final states of some reaction or decay process. In an interaction representation, for example, U can be written

$$U(t_1 t_0) = \exp\left[-(i/\hbar)\int_{t_0}^{t_1} H'dt\right]$$

where H' is the perturbing interaction.

The important thing for our purpose is that for isolated systems H', and thus U or S, is scalar under spatial rotations (conservation of total angular momentum). We establish from this that the statistical tensors ρ_{KQ} transform in the same way (6.69) as $\boldsymbol{\rho}$, and that *the tensor rank is conserved*,

$$\rho_{KQ}(\beta J, \beta' J'; t_1) = \sum_{\alpha\alpha'} \rho_{KQ}(\alpha J, \alpha' J'; t_0) U^J_{\beta\alpha}(t_1 t_0) U^{J'}_{\beta'\alpha'}(t_1 t_0)^*. \quad (6.70)$$

This means that the polarization or angular complexity of the final system (measured by the maximum rank of tensor required for its description) can never exceed that of the initial. For example a nuclear reaction initiated by s-waves will always display an isotropic angular distribution unless the colliding particles are polarized [19].

Perturbation theory is often used to calculate transition probabilities, in angular correlation problems for example [8]. This corresponds to an expansion of U or S in powers of the perturbation H', so that the spherical symmetry remains and the transformation (6.70) still holds.

CHAPTER VII

GRAPHICAL METHODS IN ANGULAR MOMENTUM

In any angular momentum coupling problem it is necessary to evaluate expressions containing sums of products of Clebsch–Gordan coefficients or Wigner $3j$-coefficients. Such calculations may often be simplified by using the graphical methods of Levinson [85]. These methods have been extended by a number of authors and the results are collected in two books by Yutsis, Levinson, and Vanagas [86] (this reference is denoted by YLV) and by Yutsis and Bandzaitis [87]. The present chapter gives a brief account of the graphical method which is complete enough to be used for solving simple angular momentum coupling problems.

The rules presented in this chapter for constructing and manipulating graphs are not identical to those given by YLV, but are related to them in a simple and well defined way (cf. section 7.2.1). Our rules are somewhat more flexible than those of YLV; they may be used to evaluate algebraic expressions involving both $3j$-symbols and Clebsch–Gordan coefficients, while the YLV rules can be used only for formulae expressed in terms of $3j$-symbols. The graphical methods of YLV are similar to those introduced by Edmonds [22] and Judd [84], but the former give the correct signs as well as the magnitudes of expressions, while the latter give only the magnitudes. Extensions of the graphical method to include tensor operators and rotations matrices have been given by several authors [82, 83].

7.1. The Basic Components of the Graphical Representation

A graphical representation is a correspondence between diagrams and algebraic formulae. Each term in an algebraic formula is represented by a component of an appropriate graph. In a consistent graphical representation it must be possible to write down the algebraic formula corresponding to a given diagram in a unique, unambiguous way.

The basic components of our graphical representation are as follows.

(1) The Wigner $3j$-symbol is represented by a node or vertex with three lines joined to it. These lines stand for the angular momenta which are coupled by the $3j$-symbol.

$$\begin{pmatrix} a & b & c \\ \alpha & \beta & \gamma \end{pmatrix} = \quad (7.1)$$

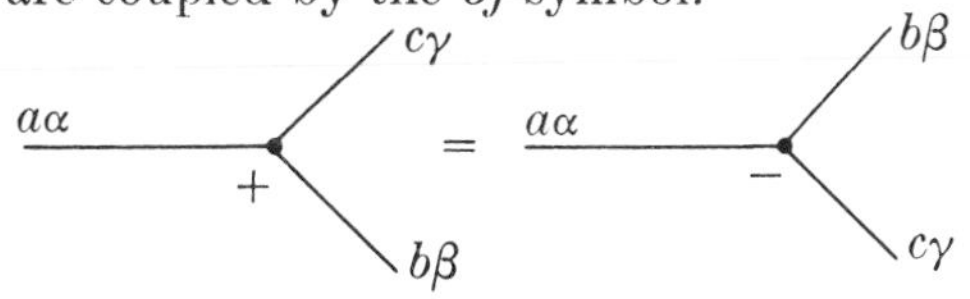

It is convenient to denote the orientation of the node by a sign; an anti-clockwise orientation is denoted by a + sign and a clockwise orientation by a − sign. Rotating a diagram does not change the cyclic order of the lines. The $3j$-coefficient has simple symmetry properties (cf. Appendix I) and remains unchanged by a cyclic permutation of the columns in the symbol. Therefore, a rotated diagram represents the same $3j$-symbol as the original diagram. The angles between the lines and the lengths of the lines have no significance. Consequently, any geometrical deformation of the diagram which preserves the orientation of the node does not change the $3j$-symbol represented by the diagram. A deformation which changes the cyclic order changes the orientation of the node and if the deformed diagram is to represent the same $3j$-symbol then the sign of the node must be changed. The symmetry relation

$$\begin{pmatrix} a & b & c \\ \alpha & \beta & \gamma \end{pmatrix} = (-)^{a+b+c}\begin{pmatrix} a & c & b \\ \alpha & \gamma & \beta \end{pmatrix}$$

implies that

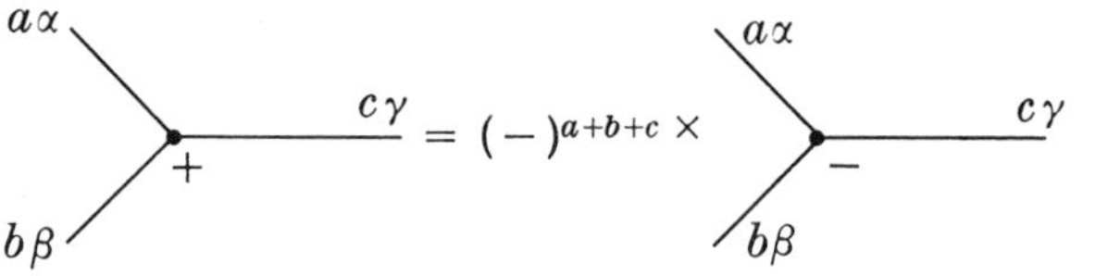

$a\alpha$, $b\beta$, $c\gamma$ (node $+$) $= (-)^{a+b+c} \times$ $a\alpha$, $b\beta$, $c\gamma$ (node $-$) (7.2)

(2) The anti-symmetric symbol or 'metric tensor'

$$\begin{pmatrix} a \\ \alpha\beta \end{pmatrix} = \delta(\alpha, -\beta)(-)^{a+\alpha}$$

which carries a phase present in many angular momentum summations is denoted by a line with an arrow on it:

$a\alpha$ ←── $b\beta$ $= \delta(ab)\begin{pmatrix} a \\ \alpha\beta \end{pmatrix}$

In particular (7.3)

$a\alpha$ ←── $a, -\alpha$ $= (-)^{a+\alpha}$ $a\alpha$ ──→ $a, -\alpha$ $= (-)^{a-\alpha}$

The $3j$-coefficient reduces to the anti-symmetric symbol when one of the angular momenta is zero:

$$\begin{pmatrix} a & b & 0 \\ \alpha & \beta & 0 \end{pmatrix} = (2a+1)^{-\frac{1}{2}}\,\delta(ab)\begin{pmatrix} a \\ \beta\alpha \end{pmatrix}. \tag{7.4}$$

This relation is represented graphically by

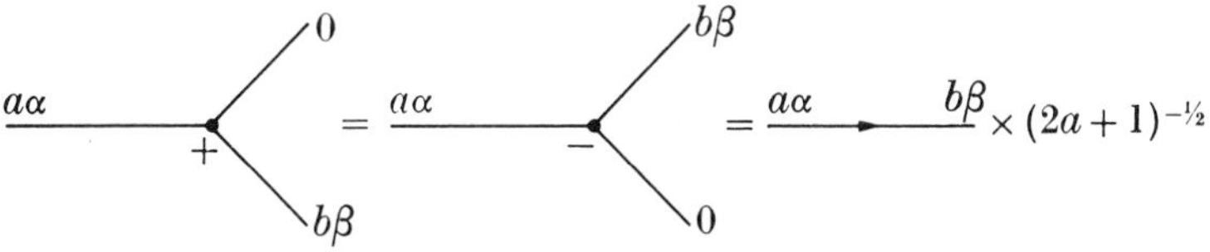

$a\alpha$ — $+$ node — 0, $b\beta$ $=$ $a\alpha$ — $-$ node — $b\beta$, 0 $=$ $a\alpha$ → $b\beta$ $\times (2a+1)^{-\frac{1}{2}}$

(7.5)

(3) An undirected line (a line with no arrow) represents the expression $\delta(a\,b)\,\delta(\alpha\,\beta)$,

$$\underline{a\alpha \qquad b\beta} = \delta(ab)\,\delta(\alpha\beta). \tag{7.6}$$

(4) More complicated diagrams may be constructed by joining the three basic components together. Two lines representing the same total angular momentum can be joined. Joining two lines implies that the z-components of the two angular momenta should be set equal and summed over. It is not necessary to write these z-components explicitly in the diagrams and we will omit them. Often we will not even write the z-component of an angular momentum corresponding to a free end of a line explicitly, and will assume that the Roman letters a, b, c ... denoting total angular momenta have z-components denoted by the corresponding Greek letters α, β, γ ... .

Lines which join nodes are called *internal lines*. *External lines* have one end connected to a node and one end free. *Closed diagrams* have no external lines. We will illustrate these definitions by constructing a few simple diagrams.

(a) The first orthogonality relation for $3j$-symbols

$$\sum_{\alpha\beta}\begin{pmatrix} a & b & c \\ \alpha & \beta & \gamma \end{pmatrix}\begin{pmatrix} a & b & c' \\ \alpha & \beta & \gamma' \end{pmatrix} = \frac{1}{(2c+1)}\,\delta(cc')\,\delta(\gamma\gamma') \tag{7.7}$$

is represented by the graphical equation

$$c \quad \overset{b}{\underset{a}{+\bigcirc-}} \quad c' \;=\; \frac{1}{(2c+1)} \times \; c \,\text{———}\, c' \tag{7.8}$$

The cyclic order of angular momenta $(a\ b\ c)$ is the same in both $3j$-coefficients in equation (7.6). This order corresponds to an anti-clockwise orientation of the first node and a clockwise orientation of the second. Hence the first node has a positive sign, while the second has a negative sign.

If we put $\gamma = \gamma'$ in equation (7.6) and sum over γ we get the result

$$-\overset{b}{\underset{a}{\ominus}}{}^{c}+ \;=1 \tag{7.9}$$

The signs of the nodes have been changed because their orientations have been reversed. The factor $(2c+1)^{-1}$ has cancelled because the summation over γ contains $(2c+1)$ equal terms.

The second orthogonality relation for $3j$-coefficients

$$\sum_{c}(2c+1)\begin{pmatrix} a & b & c \\ \alpha & \beta & \gamma \end{pmatrix}\begin{pmatrix} a & b & c \\ \alpha' & \beta' & \gamma \end{pmatrix} = \delta(\alpha\alpha')\,\delta(\beta\beta') \tag{7.10}$$

is represented by the graphical equation

$$\sum_{c}(2c+1) \quad {}^{a\alpha}_{b\beta}\!\succ\!\underset{+\qquad -}{\overset{c}{\text{———}}}\!\prec{}^{a\alpha'}_{b\beta'} \;=\; \frac{a\alpha \text{———} a\alpha'}{b\beta \text{———} b\beta'} \tag{7.11}$$

(b) The contraction of a $3j$-coefficient with an anti-symmetric symbol is represented by a node with one arrow

$$(-)^{c+\gamma}\begin{pmatrix} a & b & c \\ \alpha & \beta & -\gamma \end{pmatrix} = \sum_{\gamma}\begin{pmatrix} a & b & c \\ \alpha & \beta & \gamma' \end{pmatrix}\begin{pmatrix} c \\ \gamma\gamma' \end{pmatrix} = \; {}^{a\alpha}_{b\beta}\!\succ_{+}\!\text{——}\!\rightarrow\!\text{——}\; c\gamma \tag{7.12}$$

This graph is very useful because it gives a way of representing a Clebsch–Gordan coefficient. This advantage is the main reason for modifying the original YLV scheme. A Clebsch–Gordan coefficient is related to the $3j$-symbol by equation (3.3)

$$\langle ab\alpha\beta|c\gamma\rangle = (-)^{a-b+\gamma}(2c+1)^{\frac{1}{2}}\begin{pmatrix} a & b & c \\ \alpha & \beta & -\gamma \end{pmatrix}. \qquad (7.13)$$

Comparing equation (7.12) with equation (7.11) we get two graphical representations of the Clebsch–Gordan coefficient:

$$\langle ab\alpha\beta|c\gamma\rangle = (-)^{a-b-c}(2c+1)^{1/2} \times \qquad (7.14)$$

$$= (-)^{2a}(2c+1)^{1/2} \times \qquad (7.15)$$

In the last equation the sign of the node has been changed. This corresponds to a change in the cyclic order of the angular momenta in the $3j$-symbol and gives a factor $(-)^{a+b+c}$ because of the symmetry relation (7.2).

(c) As a final example we give graphical representations of the Racah W-function and the Wigner $6j$-symbol.

$$W(abcd;ef) = \sum (-)^{b-\beta+c-\gamma}\begin{pmatrix} a & b & e \\ \alpha & \beta & \epsilon \end{pmatrix}\begin{pmatrix} b & d & f \\ -\beta & \delta & \phi \end{pmatrix}\begin{pmatrix} d & c & e \\ \delta & \gamma & \epsilon \end{pmatrix}\begin{pmatrix} c & a & f \\ -\gamma & \alpha & \phi \end{pmatrix}$$

$=$

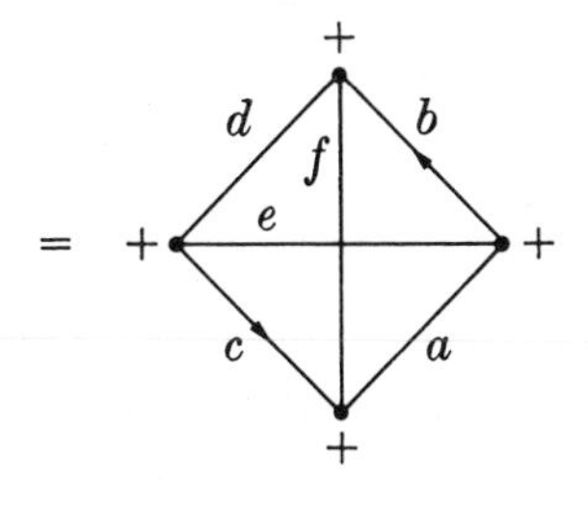

(7.16)

$$\begin{Bmatrix} a & b & e \\ d & c & f \end{Bmatrix} = \sum (-)^{a+e+c-\alpha-\epsilon-\gamma} \begin{pmatrix} a & f & c \\ \alpha & \phi & -\gamma \end{pmatrix} \begin{pmatrix} c & d & e \\ \gamma & \delta & -\epsilon \end{pmatrix} \begin{pmatrix} e & b & a \\ \epsilon & \beta & -\alpha \end{pmatrix} \begin{pmatrix} b & d & f \\ \beta & \delta & \phi \end{pmatrix}$$

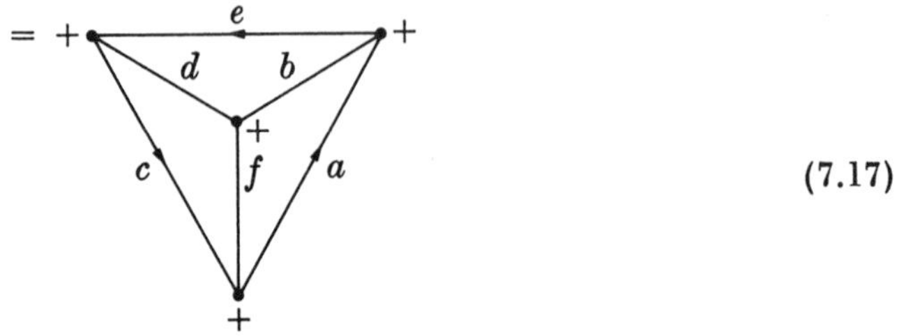

(7.17)

The sums in equations (7.16) and (7.17) are to be taken over *all* magnetic quantum numbers. Graphical representations of the 9-j symbol are given at the end of the chapter.

7.2. Simple Rules for Transforming Graphs

A calculation using the graphical technique often contains the following steps.

(i) An investigation of a physical problem leads to a formula involving sums over Clebsch–Gordan or $3j$-symbols.

(ii) The formula is represented by a graph using the rules defined in section 7.1. (Sometimes it may even be possible to go directly from the physical problem to the graph without writing down the algebraic formula.)

(iii) The graph is transformed using rules derived in this section and in section 7.3. Each transformation of the graph corresponds uniquely to some algebraic manipulation of the formula. The aim of the transformation is to isolate various components of the graph which may be identified with standard invariant functions such as the Racah W-function, the $6j$-symbol or the $9j$-symbol.

(iv) The transformed graph is reconverted to an algebraic formula. This can be done by comparison with standard graphs for $3nj$-symbols. Often the last manipulations of a graph involve adding or removing arrows or deforming the diagram to get it into some standard form. These operations are discussed in the present section.

The rules for adding or removing arrows in a graph are:

A. A line with two oppositely directed arrows is equivalent to a line with no arrows. This is a graphical expression of the result.

$$\sum_\beta \begin{pmatrix} a \\ \alpha'\beta \end{pmatrix} \begin{pmatrix} a \\ \alpha\beta \end{pmatrix} = \sum_\beta (-)^{a+\alpha+a+\alpha'}\, \delta(\alpha,-\beta)\, \delta(\alpha',-\beta) = \delta(\alpha\alpha')$$

or (7.18)

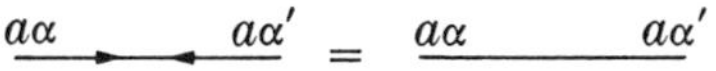

$$\underset{\longrightarrow\longleftarrow}{a\alpha \qquad a\alpha'} = \underline{a\alpha \qquad a\alpha'}$$

B. A line corresponding to an angular momentum a with two arrows in the same direction is equivalent to a line with no arrows times a factor $(-)^{2a}$.

$$\sum_\beta \begin{pmatrix} a \\ \alpha'\beta \end{pmatrix} \begin{pmatrix} a \\ \beta\alpha \end{pmatrix} = \sum_\beta (-)^{a-\alpha+a-\beta}\, \delta(\alpha-\beta)\, \delta(\beta-\alpha') = (-)^{2a}\, \delta(\alpha\alpha')$$

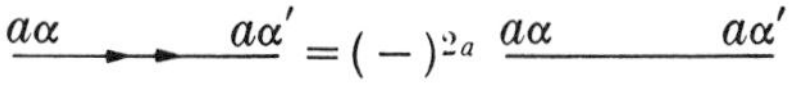

$$\underset{\longrightarrow\longrightarrow}{a\alpha \qquad a\alpha'} = (-)^{2a}\, \underline{a\alpha \qquad a\alpha'} \tag{7.19}$$

C. If an arrow on a line with angular momentum a is reversed the graph must be multiplied by a factor $(-)^{2a}$. This result follows from the symmetry relation

$$\begin{pmatrix} a \\ \alpha\alpha' \end{pmatrix} = (-)^{2a} \begin{pmatrix} a \\ \alpha'\alpha \end{pmatrix},$$

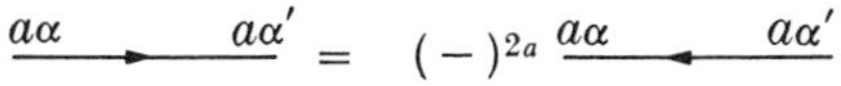

$$\underset{\longrightarrow}{a\alpha \qquad a\alpha'} = (-)^{2a}\, \underset{\longleftarrow}{a\alpha \qquad a\alpha'} \tag{7.20}$$

D. Three arrows may be added at a node, one to each line joined to the node, without changing the value of the graph provided the arrows are directed either all away from or all towards the node.

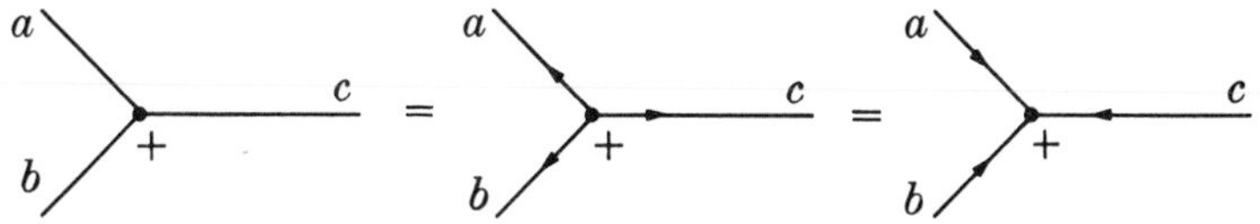

a, b, c (node +) = a, b, c (node +, arrows away) = a, b, c (node +, arrows towards)

(7.21)

The proof of the first of these relations is

$$\sum_{\alpha'\beta'\gamma'} \begin{pmatrix} a & b & c \\ \alpha' & \beta' & \gamma' \end{pmatrix} \begin{pmatrix} a \\ \alpha\alpha' \end{pmatrix} \begin{pmatrix} b \\ \beta\beta' \end{pmatrix} \begin{pmatrix} c \\ \gamma\gamma' \end{pmatrix} = (-)^{a+b+c+\alpha+\beta+\gamma} \begin{pmatrix} a & b & c \\ -\alpha & -\beta & -\gamma \end{pmatrix}$$

$$= \begin{pmatrix} a & b & c \\ \alpha & \beta & \gamma \end{pmatrix}$$

because $\alpha+\beta+\gamma = 0$ and the $3j$-coefficient has the symmetry property

$$\begin{pmatrix} a & b & c \\ -\alpha & -\beta & -\gamma \end{pmatrix} = (-)^{a+b+c} \begin{pmatrix} a & b & c \\ \alpha & \beta & \gamma \end{pmatrix}. \tag{7.22}$$

The second relation may be proved from the first and rule C.

The number and orientation of arrows in a graph may be changed in many ways using rules A–D. A graph is in *normal form* if there is exactly one arrow on every internal line. YLV have shown that only those diagrams which can be put into normal form represent formulae arising from coupling of angular momenta. A diagram with no external lines represents an invariant $3nj$-symbol *only* if it can be put into normal form.

The normal form of a graph is not unique as the directions of arrows may be changed in many ways without altering the value of the graph.

E. The direction of all arrows and the signs of all nodes may be changed simultaneously in a closed diagram without altering the value of the diagram. Let J be the sum of the total angular momenta of the internal lines. Reversing the direction of all arrows in the diagram gives a factor $(-)^{2J}$. This result follows from rule C if the graph is in normal form, because then there is exactly one arrow on every line. Adding or removing arrows by rules A, B, or D does not change the value of the graph, hence it holds for any graph which can be put into normal form. Reversing the sign at a vertex $(a\,b\,c)$ gives a factor $(-)^{a+b+c}$ (equation (7.15)). In a closed diagram each line is connected to two vertices. Hence changing the sign of every vertex produces another factor $(-)^{2J}$. This cancels the factor $(-)^{2J}$ coming from reversing the direction of all arrows.

A graph may be deformed in any way without altering its value provided

(i) the direction of any arrow relative to the nodes it connects is not changed,
(ii) the sign of a node is changed if the cyclic order of the angular momenta at the node is reversed.

7.2.1. *Relation with the Graphs of Yutsis, Levinson, and Vanagas*

The YLV graphs have an arrow on every line. A *closed* diagram in normal form of the type discussed in this chapter is completely equivalent to a YLV graph. Geometrically similar closed graphs in the two schemes represent the same algebraic formula.

A YLV graph with external lines can be converted to a graph of the type discussed here if

(i) any internal line of the YLV graph is left unchanged,
(ii) an external line with an arrow directed out of the YLV graph is replaced by a line with no arrow,
(iii) an external line $(a\alpha)$ with an arrow directed into the YLV diagram is retained with the arrow and the graph is multiplied by a factor $(-)^{a-\alpha}$.

The graphical method may be used to simplify angular momentum formulae at two different levels:

(*a*) if care is taken with the directions of arrows and the signs of nodes, the method gives the correct magnitude and sign of the result,
(*b*) if reductions are made without worrying about signs of nodes or directions of arrows the graphical method gives the correct magnitudes. In some problems where the sign of the result is not important or where it can be determined by physical arguments the magnitude of the result is all that is required.

It should be emphasized that any calculation made using graphical methods can also be made using conventional algebraic techniques. To every graphical reduction there is a corresponding algebraic reduction because of the correspondence between graphs and algebraic formulae. The graphical method has two advantages over the algebraic method:

(i) the notation is more compact because the redundant magnetic quantum numbers need not be written explicitly, and
(ii) reductions can be made by recognizing geometrical patterns.

We conclude this section with some illustrations of applications of the results presented so far.

Example 1

Verify that the graphical representation (7.16) of the Racah W-function is equivalent to the definition in terms of Clebsch–Gordan coefficients given in equation (3.13).

If equation (3.13) is summed over the magnetic quantum number γ then we get

$$\{(2e+1)(2f+1)(2c+1)^2\}^{\frac{1}{2}}\, W(abcd,ef) = \sum \langle ab\beta\alpha|e\epsilon\rangle\langle ed\epsilon\delta|c\gamma\rangle\langle bd\beta\delta|f\phi\rangle\langle af\alpha\phi|c\gamma\rangle,$$

where the sum is taken over all magnetic quantum numbers. We write this equation as a graph using the representation (7.18) for each Clebsch–Gordan coefficient.

$\{(2e+1)(2f+1)\}^{\frac{1}{2}}$
$(2c+1)W(abcd;ef) =$

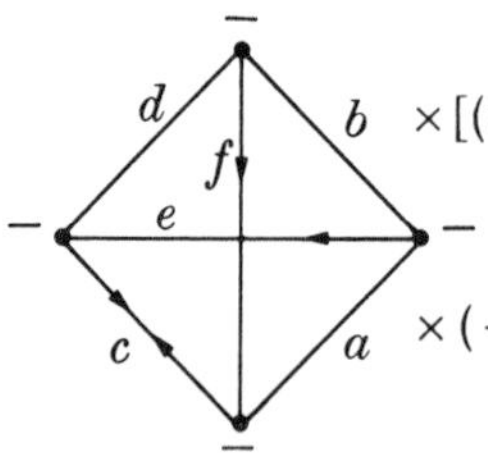

$\times[(2e+1)(2f+1)(2c+1)^2]^{\frac{1}{2}}$
$\times(-)^{2a+2e+2b+2a}$

(7.23)

The diagram in (7.23) can be reduced by omitting the oppositely directed arrows on the line c (rule A). Reversing the arrow on line e and inserting two arrows in the same direction on line b

cancels the factor $(-)^{2e+2b}$. Hence we get after cancelling the square roots

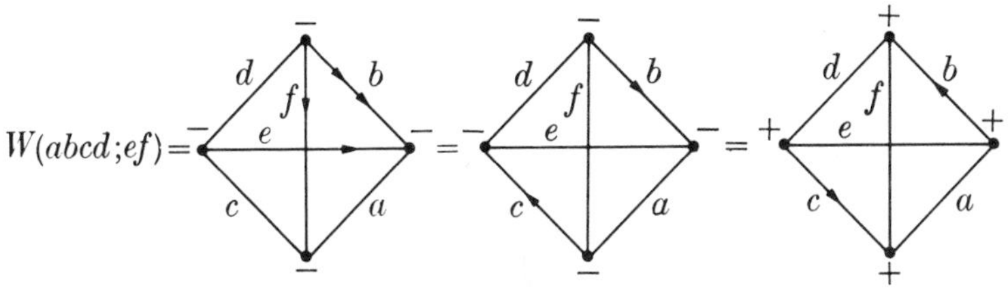

$$W(abcd;ef) = \text{[graph]} = \text{[graph]} = \text{[graph]} \tag{7.24}$$

In the first step of the reduction in equation (7.24) arrows directed out of the vertices (abe) and (afc) are inserted using rule D, and then pairs of oppositely directed arrows are removed from the lines a,b,e,f by rule A. These operations leave arrows directed in a clockwise sense on lines c and b and negative signs on the vertices. Finally the signs of all the vertices and the directions of all arrows are reversed using rule E to obtain the graph of the W-function in equation (7.16).

Example 2

Prove the symmetry relation

$$W(abcd;ef) = (-)^{b+c-e-f}\, W(aefd;bc).$$

First draw the graph for the W-function given in equation (7.16)

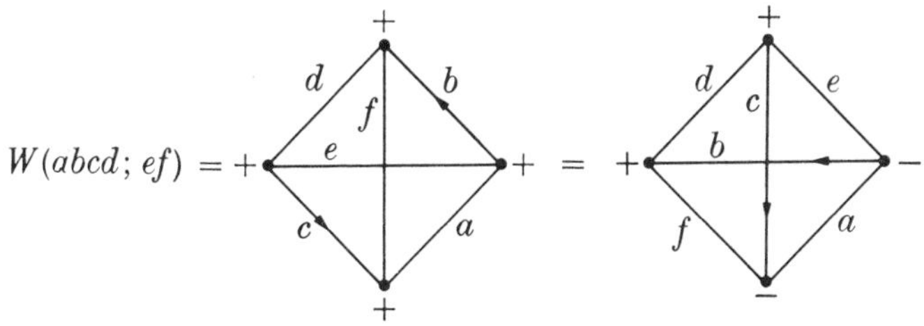

$$W(abcd;\, ef) = \text{[graph]} = \text{[graph]} \tag{7.25}$$

The diagram has been deformed so that e and f form two opposite edges. The deformation changes the cyclic order at the nodes (afc) and (abe) so the signs of those nodes must be reversed. The arrows may be transferred from the lines b and c to the lines e and f by inserting arrows directed into the node (afc) and out of the node (aeb). Pairs of arrows in the same direction on lines a, b, and c can then be removed giving a factor $(-)^{2a+2b+2c}$. Hence

$$W(abcd;\, ef) = (-)^{2a+2b+2c} \times$$

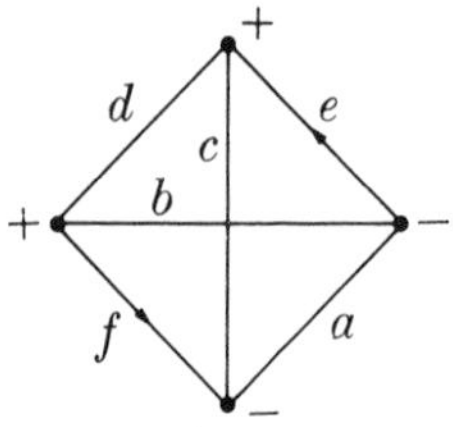

$$= (-)^{b+c-e-f}$$

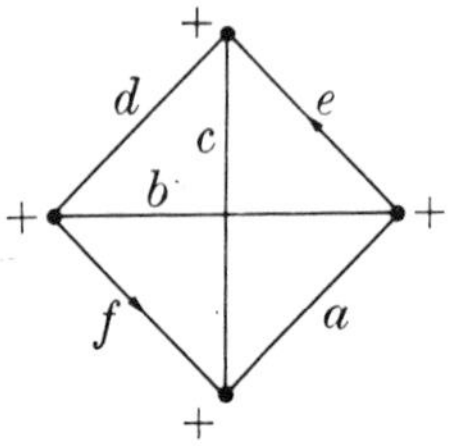

In the final step the signs of the vertices (afc) and (abe) have been reversed giving an additional factor $(-)^{(a+f+c)+(a+b+e)}$ and the desired result obtained by comparison of the last diagram with equation (7.16).

Example 3

Prove

$$W(abcd;ef) = (-)^{a+b+c+d} \begin{Bmatrix} a & b & e \\ d & c & f \end{Bmatrix}$$

where the W-function and the $6j$-symbol are defined in equations (7.16) and (7.17).

The result can be proved by the following sequence of graphical transformations.

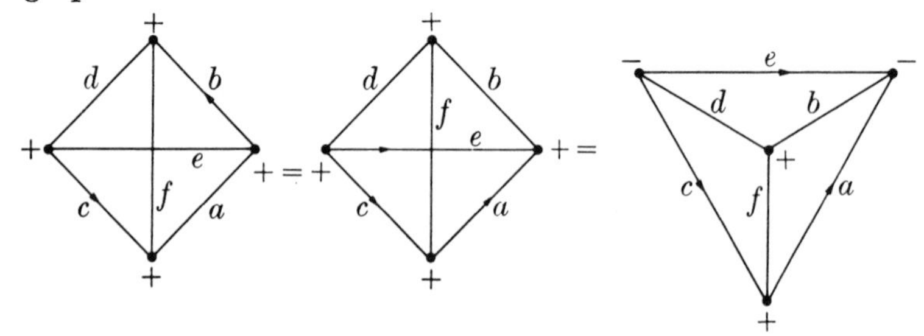

$= (-)^{a+b+c+d}$ 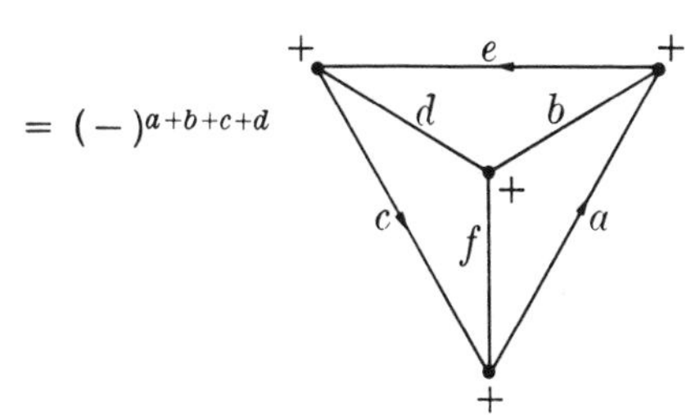

A phase $(-)^{a+b+e+c+d+e}$ comes from reversing the signs of the nodes (abe) and (cde) and a factor $(-)^{2e}$ from changing the direction of the arrow on the line e.

Example 4

The following graphs all represent the same 9-j symbol:

$$\begin{Bmatrix} a & b & c \\ d & e & f \\ g & h & i \end{Bmatrix}$$

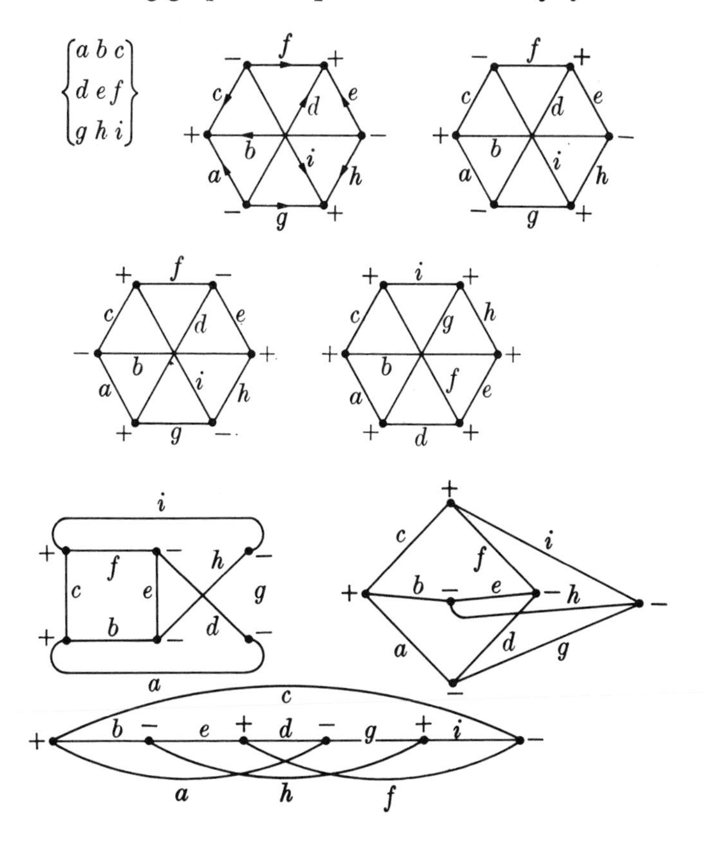

7.3. Theorems on Block Diagrams

This section begins with a discussion of certain generalized angular momentum coupling coefficients constructed from $3j$-coefficients and anti-symmetric symbols which have been called jm-coefficients by Yutsis, Levinson, and Vanagas (YLV). Let $\phi_1(j_1m_i), \ldots \phi_n(j_nm_n)$ be the wave functions of the components of a quantum mechanical system. The function $F_n\begin{pmatrix} j_1 & \cdots & j_n \\ m_1 & \ldots & m_n \end{pmatrix}$ is called a jm-coefficient if it couples the angular momenta of the states $\phi_1 \ldots \phi_n$ to a zero resultant, that is if the state

$$\Phi = \sum_{m_i} F_n\begin{pmatrix} j_1 & \cdots & j_n \\ m_1 & \ldots & m_n \end{pmatrix} \phi_1(j_1m_1) \ldots \phi_n(j_nm_n) \quad (7.26)$$

is a scalar invariant. In general F_n will be a sum of products of $3j$-coefficients and anti-symmetric symbols which would be represented graphically by a diagram with n external lines. The detailed internal structure of the graph is of no importance for the questions considered in this section, and it is convenient to denote the graph of F_n by a block with n external lines

$$F_n\begin{pmatrix} j_1 \cdots j_n \\ m_1 \ldots m_n \end{pmatrix} =$$

$j_n m_n$

$j_1 m_1$

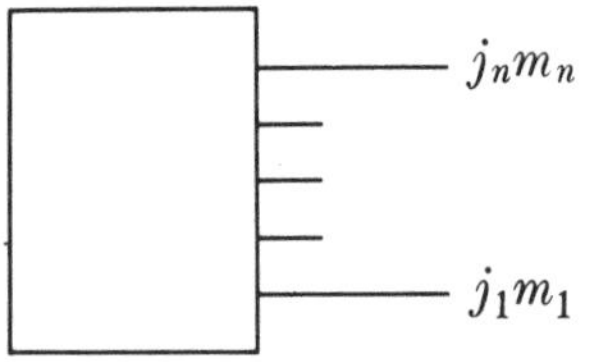

(7.27)

Conditions which an expression F_n must satisfy in order that it should be a jm-coefficient have been investigated by YLV. They may be stated in terms of the graphical representation presented in this chapter in the following way.

The expression F_n is a jm-coefficient if, by using rules A–D of section 7.2, its graph can be put into a form in which every internal line has exactly one arrow on it and every external line has no arrow on it.

The anti-symmetric symbol $\begin{pmatrix} a \\ \alpha\beta \end{pmatrix}$ and the $3j$-symbol $\begin{pmatrix} abc \\ \alpha\beta\gamma \end{pmatrix}$

are the simplest examples of *jm*-coefficients. General *jm*-coefficients retain some of the properties of these simple functions. For example

(i) The *jm*-coefficient

$$F_n\begin{pmatrix} j_1 & \dots & j_n \\ m_1 & \dots & m_n \end{pmatrix} = 0 \tag{7.28}$$

unless $\sum_1^n j_i$ is integral and $\sum_1^n m_i = 0$.

(ii)

$$F_n\begin{pmatrix} j_1 & \dots & j_n \\ -m_1 & \dots & -m_n \end{pmatrix} = (-)^{j_1+\dots+j_n} F_n\begin{pmatrix} j_1 & \dots & j_n \\ m_1 & \dots & m_n \end{pmatrix}. \tag{7.29}$$

The following theorems hold for *jm*-coefficients F_n with $n = 1$, 2, and 3.

THEOREM I. *If* $F_1\begin{pmatrix} j \\ m \end{pmatrix}$ *is a jm-coefficient associated with a graph with one external line then the function is zero unless* $j = m = 0$.

$$F_1\begin{pmatrix} j \\ m \end{pmatrix} = F_1\begin{pmatrix} 0 \\ 0 \end{pmatrix} \delta(j0)\, \delta(m0). \tag{7.30}$$

THEOREM II. *If* $F_2\begin{pmatrix} j_1\, j_2 \\ m_1 m_2 \end{pmatrix}$ *is a jm-coefficient associated with a graph with two external lines then*

$$F_2\begin{pmatrix} j_1\, j_2 \\ m_1 m_2 \end{pmatrix} = \frac{1}{(2j_1+1)} \delta(j_1 j_2) \begin{pmatrix} j_1 \\ m_1 m_2 \end{pmatrix} \bar{F}_2 \tag{7.31}$$

where $$\bar{F}_1 = \sum_{m_1 m_2} \begin{pmatrix} j_1 \\ m_1 m_2 \end{pmatrix} \times F_2\begin{pmatrix} j_1\, j_1 \\ m_1 m_2 \end{pmatrix}.$$

THEOREM III. *If* $F_3\begin{pmatrix} j_1\, j_2\, j_3 \\ m_1 m_2 m_3 \end{pmatrix}$ *is a jm-coefficient associated with a graph with three external lines then*

$$F_3\begin{pmatrix} j_1\, j_2\, j_3 \\ m_1 m_2 m_3 \end{pmatrix} = \begin{pmatrix} j_1\, j_2\, j_3 \\ m_1 m_2 m_3 \end{pmatrix} \times \bar{F}_3 \tag{7.32}$$

where $$\bar{F}_3 = \sum_{m_1 m_2 m_3} \begin{pmatrix} j_1\, j_2\, j_3 \\ m_1 m_2 m_3 \end{pmatrix} \times F_3\begin{pmatrix} j_1\, j_2\, j_3 \\ m_1 m_2 m_3 \end{pmatrix}.$$

The proof of Theorem I follows directly from the definition (7.26) of a jm-coefficient. The only case in which one angular momentum (jm) can be coupled to give a zero angular momentum occurs if $j = 0$. Theorem II can be reduced to a special case of Theorem I by coupling the angular momenta $(j_1 m_1)$ and $(j_2 m_2)$ to a resultant (jm). Graphically F_2 is represented by a block with two external lines. If these lines are coupled using the relation (7.11) we get

$$F_2\begin{pmatrix} j_1 & j_2 \\ m_1 & m_2 \end{pmatrix} = \sum_j (2j+1)$$

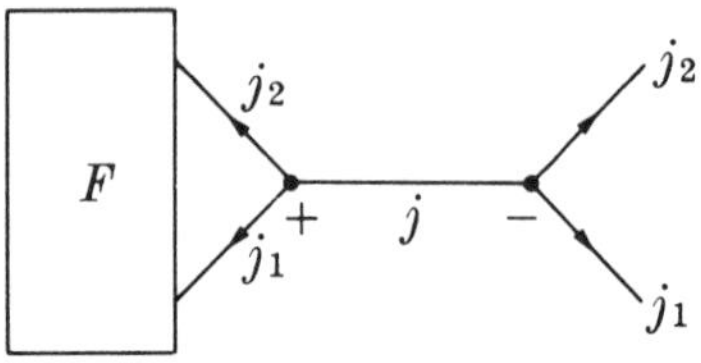

F j_2 j_1 + j − j_2 j_1

(Arrows have been added at the vertices by rule D and removed from the line j by rule A in order to get the left-hand part of the graph in normal form.) By Theorem I this graph is zero unless $j = 0$. Hence, using relation (7.5) and cancelling redundant arrows,

$$F_2\begin{pmatrix} j_1 & j_2 \\ m_1 & m_2 \end{pmatrix} = \frac{\delta(j_1 j_2)}{(2j_1+1)}$$

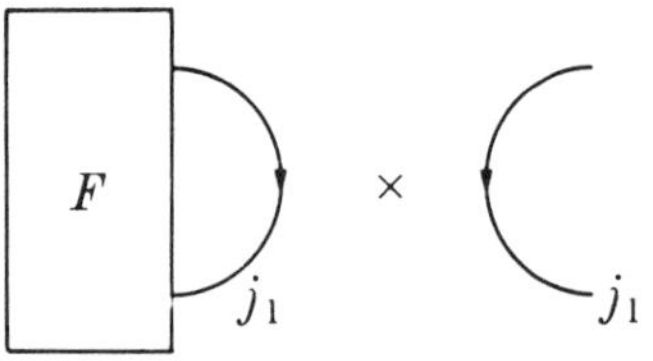

F j_1 × j_1

which is the graphical form of equation (7.31). Theorem III can be proved by reducing it to a special case of Theorem II in a similar way. Details of this proof will be left as a problem for the reader.

The properties of jm-coefficients discussed in the first part of this section may sometimes be used to decompose a complicated graph into simpler components. Suppose a graph can be separated into two blocks F and G where F has either one, two, or three external lines and is joined to G by these lines. It follows from Theorems I, II, and III of this section that such a graph will split up into a product of disconnected components, one associated with the block F and the other with G. If the graph has one external line and is connected to G by this line, then Theorem I shows that the angular momentum associated with this line must be zero. The graphs F and G are disconnected by simply omitting this line and using relation (7.4) at the associated nodes in F and G. The cases where F has two or three external lines will be considered in more detail.

Theorems II and III can be used only if the graph F represents a *jm*-coefficient. This condition is satisfied automatically if the graph F is in normal form with an arrow on every internal line and any arrows on the lines joining F to G are incorporated in G. If these conditions are satisfied, and if F and G are connected by either two or three lines, then the graphs decompose as follows.

(a) *Two connecting lines*

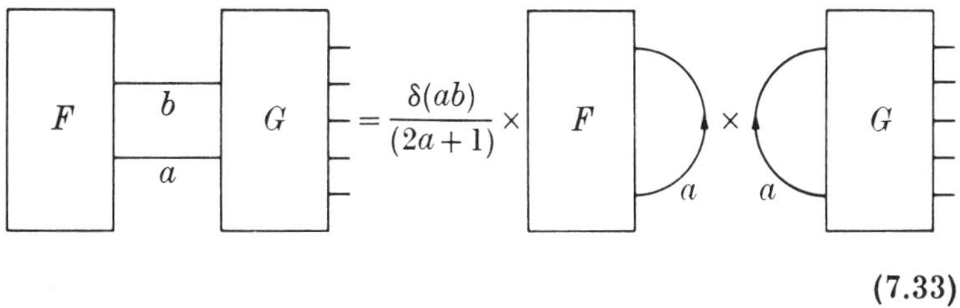

(7.33)

(b) *Three connecting lines*

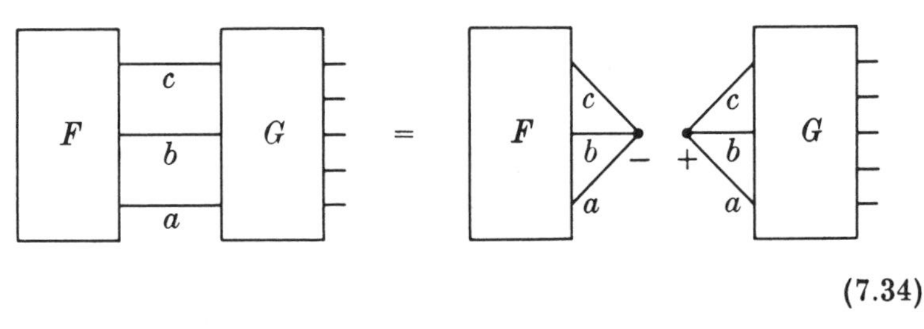

(7.34)

These results follow directly from equations (7.31) and (7.32). In applications it is not necessary to put F explicitly into normal form. It is sufficient to be certain that any graphs formed by decomposing a larger graph *can* be put into normal form by using the rules A–D. To satisfy this condition it is sometimes necessary to add pairs of arrows on the connecting lines before breaking them (cf. problem 6). We shall illustrate the decomposition (7.34) by an example.

Example 5

Prove the relation

$$\sum_{\delta\epsilon\phi}\begin{pmatrix} d & e & c \\ -\delta & \epsilon & \gamma \end{pmatrix}\begin{pmatrix} e & f & a \\ -\epsilon & \phi & \alpha \end{pmatrix}\begin{pmatrix} f & d & b \\ -\phi & \delta & \beta \end{pmatrix}(-)^{d+e+f-\delta-\epsilon-\phi} =$$

$$= \begin{Bmatrix} a & b & c \\ d & e & f \end{Bmatrix}\begin{pmatrix} a & b & c \\ \alpha & \beta & \delta \end{pmatrix}. \quad (7.35)$$

The result can be proved graphically by using (7.34) in the following way:

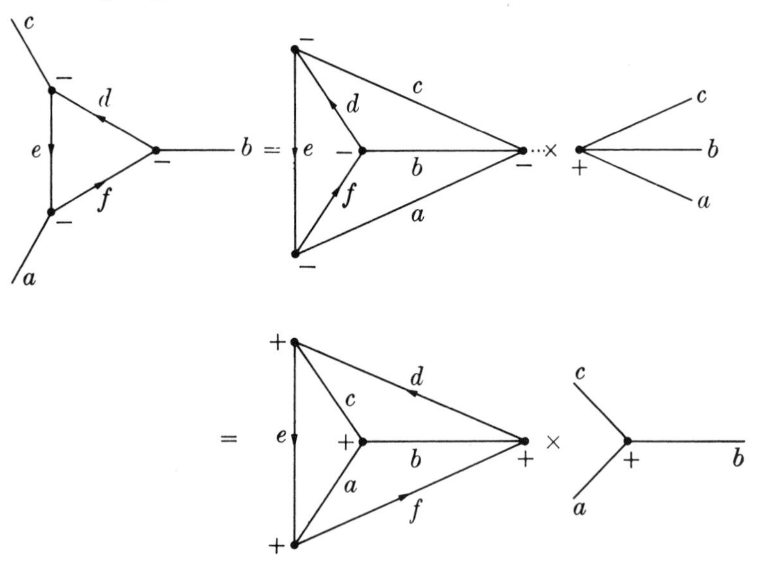

Graphs which can be separated into blocks F and G where F has three or less external lines are called reducible graphs. We have seen that such a graph splits up into a product of factors. Irreducible graphs cannot be reduced to a product of factors, but often they can be written as a sum of products of simpler components. We will discuss the most important case where a graph may be divided into blocks F_4 and G where F_4 has four external lines. The block F_4 will be assumed to have an arrow on every internal line and arrows on the four lines joining F_4 to G will be associated with G. If these conditions are satisfied, F_4 represents a jm-coefficient. The angular momenta associated with the external lines of F_4 are denoted by a, b, c, and d. The graph may be decomposed by coupling the angular momenta a and b to a resultant x and c and d to a resultant y by using the graphical relation (7.10).

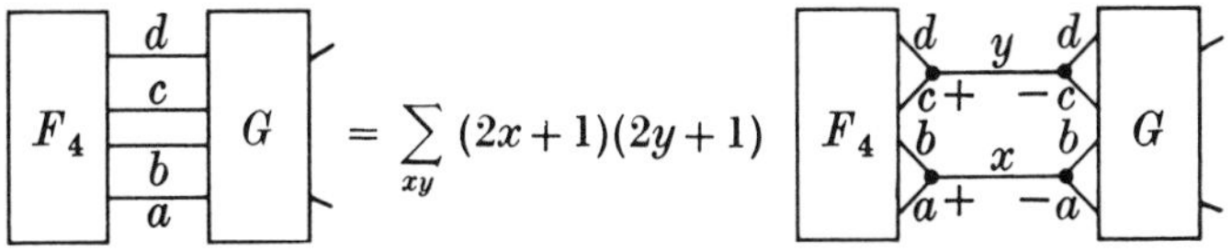

$$= \sum_{xy} (2x+1)(2y+1) \tag{7.36}$$

$$= \sum_{x} (2x+1) \tag{7.37}$$

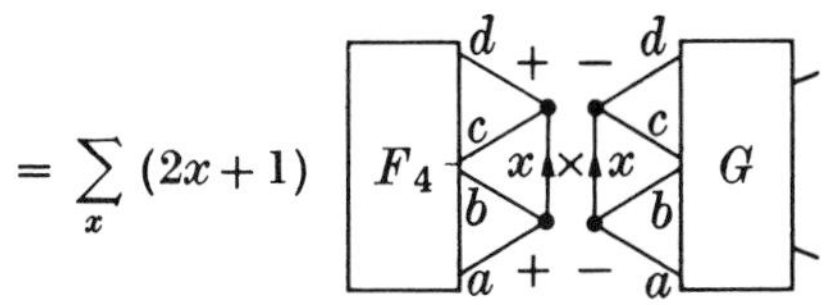

The result (7.37) follows from (7.36) by using Theorem II or the equivalent graphical decomposition (7.33). Arrows may be added to vertices using rule D of section 7.2 if desired and the signs of the nodes may be chosen in several different ways. We illustrate this result by an example:

Example 6

The jm-coefficients

$$E_e\begin{pmatrix} a & c & b & d \\ \alpha & \gamma & \beta & \delta \end{pmatrix} = \sum_{\epsilon} (-)^{e-\epsilon} \begin{pmatrix} e & a & c \\ \epsilon & \alpha & \gamma \end{pmatrix} \begin{pmatrix} e & b & d \\ -\epsilon & \beta & \delta \end{pmatrix}$$

and

$$F_f\begin{pmatrix} a & b & c & d \\ \alpha & \beta & \gamma & \delta \end{pmatrix} = \sum_{\phi} (-)^{f-\phi} \begin{pmatrix} f & a & b \\ \phi & \alpha & \beta \end{pmatrix} \begin{pmatrix} f & c & d \\ -\phi & \gamma & \delta \end{pmatrix}$$

represent two ways of coupling four angular momenta to a zero resultant. Find the relation between the two coupling schemes.

Drawing the graph for E and using the decomposition (7.37) we get

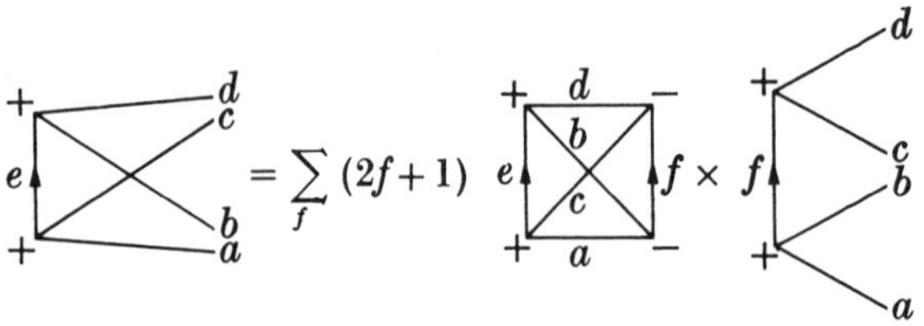

or $E_e\begin{pmatrix} a & c & b & d \\ \alpha & \gamma & \beta & \delta \end{pmatrix} = \sum_f (2f+1)(-)^{b+c+e+f}\begin{Bmatrix} a & b & f \\ d & c & e \end{Bmatrix} F_f\begin{pmatrix} a & b & c & d \\ \alpha & \beta & \gamma & \delta \end{pmatrix}.$ (7.38)

Equation (7.38) is another of the familiar relations between $6j$-coefficients and $3j$-coefficients (Appendix II).

An important special case of the result (7.37) arises when a sequence of blocks each with four external lines is connected as in the following graph.

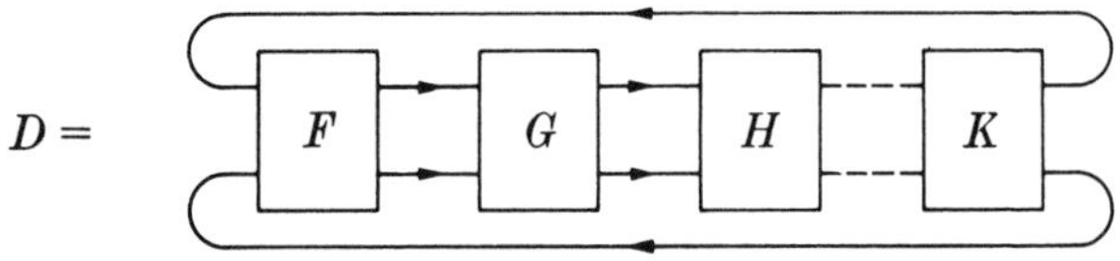

We assume that every internal line in F, G, ... has exactly one arrow on it, and that arrows on connecting lines are shown explicitly. Hence each of the blocks F, G, ... represents a jm-coefficient and the diagram D is in normal form.

This graph reduces to the following sum:

$$D = \sum_y (2y+1)F(y)G(y) \; \ldots \; K(y), \tag{7.39}$$

where

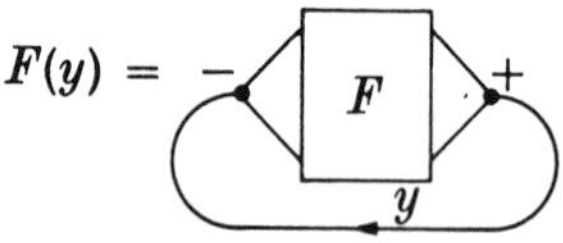

and similarly for $G(y)$, etc. The result may be proved in the same way as (7.37).

The results of this section are illustrated by a graphical proof of the Biedenharn–Elliott sum rule [5], [23].

Example 7

Consider the graph

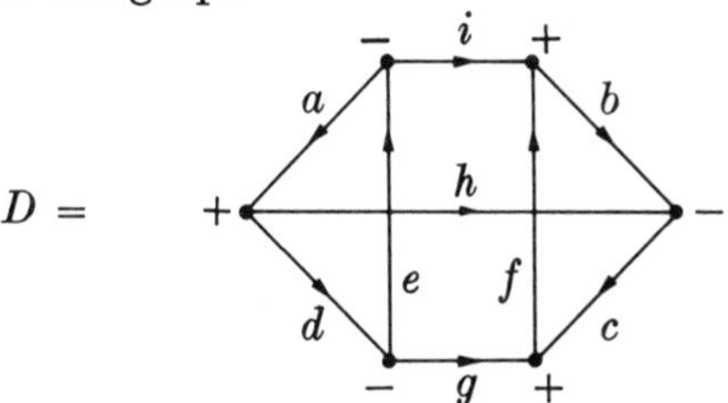

This graph may be decomposed by separating it on the lines (ghj) using the result (7.34)

$$D = \ldots \times \ldots = \begin{Bmatrix} g & h & i \\ a & e & d \end{Bmatrix} \begin{Bmatrix} g & h & i \\ b & f & c \end{Bmatrix} \tag{7.40}$$

It may also be decomposed by using the result (7.39)

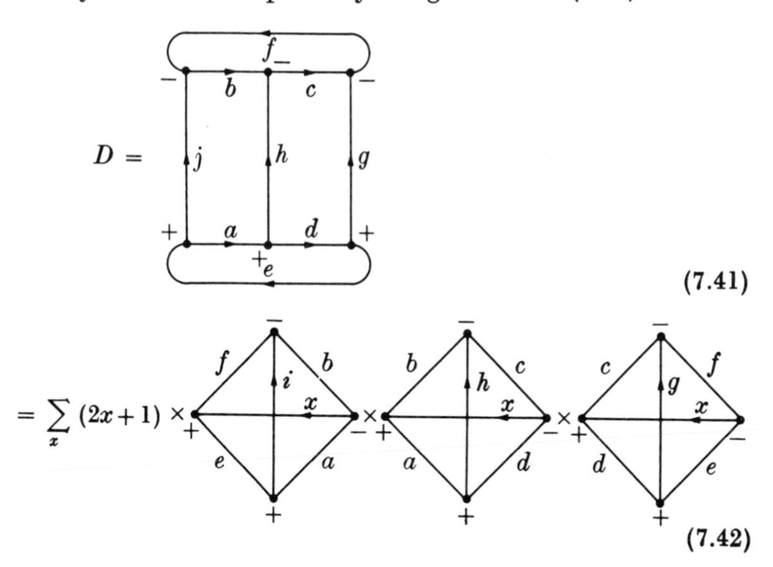

The $6j$-coefficients can be extracted from the graphs in (7.42) by comparing them with the standard graph (7.17):

$$D = \sum_{x} (2x+1)(-)^{a+b+c+d+e+f+g+h+i+x}\begin{Bmatrix} e & f & x \\ b & a & i \end{Bmatrix}\begin{Bmatrix} a & b & x \\ c & d & h \end{Bmatrix}\begin{Bmatrix} d & c & x \\ f & e & g \end{Bmatrix}. \tag{7.43}$$

For example the first graph in equation (7.42) has a value $(-)^{e+b-i-x}\begin{Bmatrix} e & f & x \\ b & a & i \end{Bmatrix}$. We use the facts that $g+h+i$ is an integer and $(-)^{-3x} = (-)^{x}$ in obtaining the final form of equation (7.43).

Example 8

Evaluate the expression

$$\begin{aligned} F(kq,k'q') = \sum (-)^{l-m+J-N} &\langle ll'-mm'|kq\rangle\langle JJ'-NN'|k'q'\rangle \\ \times\ &\langle ll_1mm_1|LM\rangle\langle l_1l_2m_1m_2|KQ\rangle\langle Ll_2Mm_2|JN\rangle \times \\ \times\ &\langle l'l_1'm'm_1'|L'M'\rangle\langle l_1'l_2'm_1'm_2'|KQ\rangle\langle L'l_2'M'm_2'|J'N'\rangle, \end{aligned}$$

where the sum is taken over all magnetic quantum numbers except for q and q'.

Using the relations (7.12) and (7.14) this expression reduces to

$$F = F_1(-)^{\phi}[(2L+1)(2L'+1)(2J+1)(2J'+1)(2k+1)(2k'+1)]^{\frac{1}{2}} \times (2K+1)$$

where

$$\phi = 2l+2L+2l_1+2l'+2L'+2l_1'+2l+2J,$$

(the phase $(-)^{\phi} = (-)^{2l+2J}$ because (lLl_1) and $(l'L'l_1')$ satisfy triangular conditions) and F_1 is represented by the graph

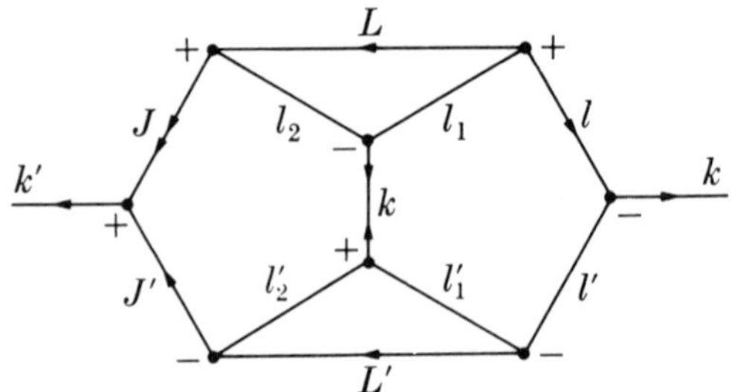

From Theorem III the graph for F_1 may be separated on the lines l, J, and K, and on the lines l', J', and K giving a product of three factors G_1, G_2, and G_3. These factors are

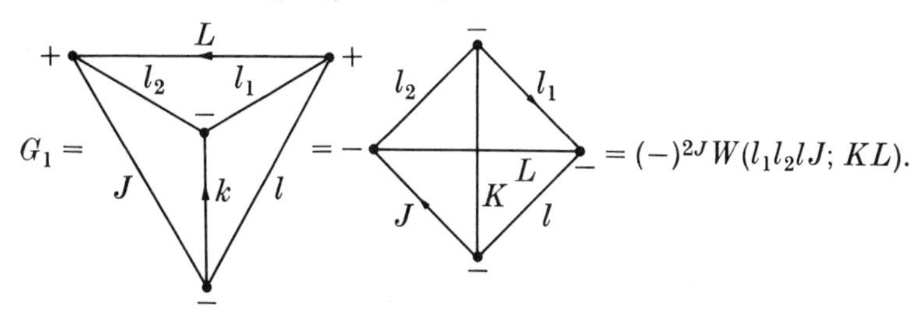

$G_1 =$

$= -$

$= (-)^{2J} W(l_1 l_2 l J;\, KL).$

$G_2 =$

$= (-)^{2J'} W(l_1' l_2' l' J';\, KL')$

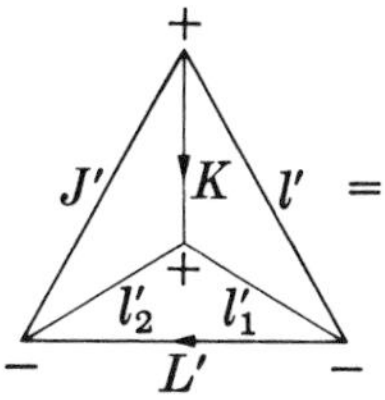

$G_3 =$

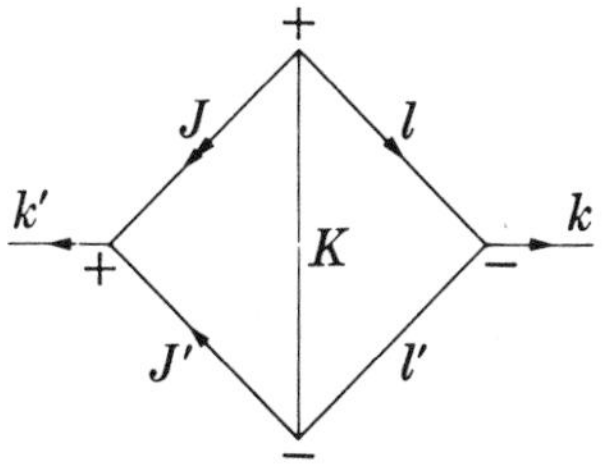

The graph for G_3 can be reduced by using Theorem II giving

$G_3 =$

$\times \dfrac{1}{2k+1}\,\delta(kk')\,\delta(qq')$

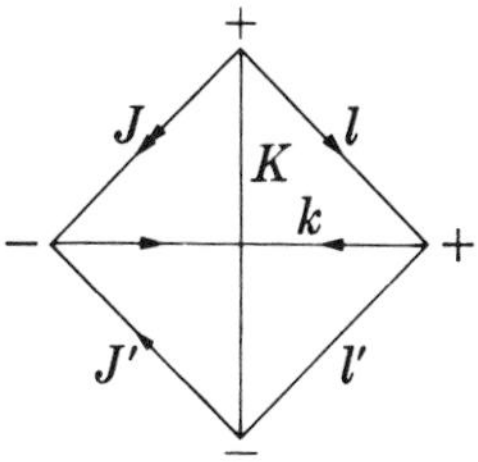

$$= \frac{(-)^{\phi_1}}{2k+1} W(ll'JJ';kK)\,\delta(kk')\,\delta(qq'),$$

where $\phi_1 = 2J-(l+J+K)+(l'+l+k) = l'+k+J-K.$

Collecting the results we get

$$F(kq,k'q') = W(l_1l_2lJ;KL)\,W(l_1'l_2'l'J';KL')\,W(ll'JJ';kK)\times$$
$$\times\,\delta(kk')\,\delta(qq')(-)^{J+l'+K-k}(2K+1)[(2L+1)(2L'+1)\times$$
$$\times\,(2J+1)(2J'+1)]^{\frac{1}{2}}.$$

APPENDIX I

3-j AND CLEBSCH–GORDAN COEFFICIENTS

THE CG coefficient is defined by the transformation (2.30)

$$|abc\gamma\rangle = \sum_{\alpha\beta} |ab\alpha\beta\rangle\langle ab\alpha\beta|c\gamma\rangle$$

and vanishes unless $\alpha+\beta = \gamma$. Other authors use the notations $(ab\alpha\beta|abc\gamma)$ [17], $(a\alpha b\beta|abc\gamma)$ [22], $C_{ab}(c\gamma;\ \alpha\beta)$ [9], $C(abc;\ \alpha\beta)$ [54], $C^{abc}_{\alpha\beta\gamma}$ [65], $C^{c\gamma}_{a\alpha b\beta}$ [39], $S^{(ab)}_{c\alpha\beta}$ [78], and $C^{c}_{\alpha\beta}$ [42] for the same quantity. The Wigner 3-j is related to it by

$$\langle ab\alpha\beta|c-\gamma\rangle = (-)^{a-b-\gamma}(2c+1)^{\frac{1}{2}}\begin{pmatrix} a & b & c\\ \alpha & \beta & \gamma\end{pmatrix}.$$

Note the appearance of γ with a minus sign on the left, so that now $a+\beta+\gamma = 0$. Related quantities are defined by Racah [31]

$$V\begin{pmatrix} a & b & c\\ \alpha & \beta & \gamma\end{pmatrix} \equiv V(abc;\ \alpha\beta\gamma) = (-)^{c+b-a}\begin{pmatrix} a & b & c\\ \alpha & \beta & \gamma\end{pmatrix}.$$

Orthogonality relations:

$$\sum_{\alpha\beta}(2c+1)\begin{pmatrix} a & b & c\\ \alpha & \beta & \gamma\end{pmatrix}\begin{pmatrix} a & b & c'\\ \alpha & \beta & \gamma'\end{pmatrix} = \delta_{cc'}\delta_{\gamma\gamma'},$$

$$\sum_{c\gamma}(2c+1)\begin{pmatrix} a & b & c\\ \alpha & \beta & \gamma\end{pmatrix}\begin{pmatrix} a & b & c\\ \alpha' & \beta' & \gamma\end{pmatrix} = \delta_{\alpha\alpha'}\delta_{\beta\beta'}.$$

Symmetry: If the 3-j is rewritten [51]

$$\begin{pmatrix} a & b & c\\ \alpha & \beta & \gamma\end{pmatrix} = \begin{bmatrix} b+c-a & c+a-b & a+b-c\\ a-\alpha & b-\beta & c-\gamma\\ a+\alpha & b+\beta & c+\gamma\end{bmatrix},$$

it is invariant under interchange of rows and columns (reflection about diagonals), and is multiplied by $(-)^{a+b+c}$ upon interchange of two adjacent rows or columns, giving 72 equivalent symbols. In particular this means

$$\begin{pmatrix} a & b & c \\ -\alpha & -\beta & -\gamma \end{pmatrix} = (-)^{a+b+c} \begin{pmatrix} a & b & c \\ \alpha & \beta & \gamma \end{pmatrix}$$

and that the 3-j is invariant under cyclic permutation of its columns and multipied by $(-)^{a+b+c}$ by non-cyclic ones

$$\begin{pmatrix} a & b & c \\ \alpha & \beta & \gamma \end{pmatrix} = \begin{pmatrix} b & c & a \\ \beta & \gamma & \alpha \end{pmatrix} = (-)^{a+b+c} \begin{pmatrix} b & a & c \\ \beta & \alpha & \gamma \end{pmatrix}, \text{ etc.}$$

Recurrence relations:

$$[(c\mp\gamma)(c\pm\gamma+1)]^{\frac{1}{2}} \begin{pmatrix} a & b & c \\ \alpha & \beta & \gamma\pm 1 \end{pmatrix} +$$

$$+[(a\mp\alpha)(a\pm\alpha+1)]^{\frac{1}{2}} \begin{pmatrix} a & b & c \\ \alpha\pm 1 & \beta & \gamma \end{pmatrix} +$$

$$+[(b\mp\beta)(b\pm\beta+1)]^{\frac{1}{2}} \begin{pmatrix} a & b & c \\ \alpha & \beta\pm 1 & \gamma \end{pmatrix} = 0,$$

$$[(a+b+c+1)(b+c-a)]^{\frac{1}{2}} \begin{pmatrix} a & b & c \\ \alpha & \beta & \gamma \end{pmatrix} = [(b+\beta)(c-\gamma)]^{\frac{1}{2}} \times$$

$$\times \begin{pmatrix} a & b-\frac{1}{2} & c-\frac{1}{2} \\ \alpha & \beta-\frac{1}{2} & \gamma+\frac{1}{2} \end{pmatrix} - [(b-\beta)(c+\gamma)]^{\frac{1}{2}} \begin{pmatrix} a & b-\frac{1}{2} & c-\frac{1}{2} \\ \alpha & \beta+\frac{1}{2} & \gamma-\frac{1}{2} \end{pmatrix},$$

$$[(a+b+c+2)(b+c-a+1)(a+c-b+1)(a+b-c)]^{\frac{1}{2}} \begin{pmatrix} a & b & c+1 \\ \alpha & \beta & \gamma \end{pmatrix}$$

$$= [(b-\beta)(b+\beta+1)(c+\gamma)(c+\gamma+1)]^{\frac{1}{2}} \begin{pmatrix} a & b & c \\ \alpha & \beta+1 & \gamma-1 \end{pmatrix}$$

$$-2\beta[(c+\gamma+1)(c-\gamma+1)]^{\frac{1}{2}} \begin{pmatrix} a & b & c \\ \alpha & \beta & \gamma \end{pmatrix} - [(b+\beta)\times$$

$$\times (b-\beta+1)(c-\gamma)(c-\gamma+1)]^{\frac{1}{2}} \begin{pmatrix} a & b & c \\ \alpha & \beta-1 & \gamma+1 \end{pmatrix}.$$

Algebraic formulae for the general 3-j are given by equation (2.34), and for $\alpha = \beta = 0$ by equation (2.35). Formulae for

$c = \frac{1}{2}$, 1, $\frac{3}{2}$, 2 are given in Table 3. Other special cases are

$$\begin{pmatrix} a & b & 0 \\ \alpha & \beta & 0 \end{pmatrix} = (-)^{a-\alpha}\delta_{ab}\delta_{\alpha,-\beta}\ (2a+1)^{-\frac{1}{2}}$$

$$\begin{pmatrix} a & b & a+b \\ \alpha & \beta & \gamma \end{pmatrix}$$

$$= (-)^{a-b-\gamma}\left[\frac{(2a)!\ (2b)!\ (a+b+\gamma)!\ (a+b-\gamma)!}{(2a+2b+1)!\ (a+\alpha)!\ (a-\alpha)!\ (b+\beta)!\ (b-\beta)!}\right]^{\frac{1}{2}},$$

$$\begin{pmatrix} a & b & a+b-1 \\ \alpha & \beta & \gamma \end{pmatrix} = (-)^{a-b-\gamma}2(b\alpha-a\beta)\times$$

$$\times\left[\frac{(2a-1)!\ (2b-1)!\ (a+b+\gamma-1)!\ (a+b-\gamma-1)!}{(2a+2b)!\ (a+\alpha)!\ (a-\alpha)!\ (b+\beta)!\ (b-\beta)!}\right]^{\frac{1}{2}},$$

$$\begin{pmatrix} a & b & a+b-2 \\ \alpha & \beta & \gamma \end{pmatrix} = (-)^{a-b-\gamma}\times$$

$$\times\left[\frac{(a+b-\gamma-2)!\ (a+b+\gamma-2)!\ (2a-2)!\ (2b-2)!}{2(a-\alpha)!\ (a+\alpha)!\ (b+\beta)!\ (b-\beta)!\ (2a+2b-1)!}\right]^{\frac{1}{2}}\times$$

$$\times[a+\alpha)(a+\alpha-1)(b-\beta)(b-\beta-1)+(a-\alpha)(a-\alpha-1)\times$$

$$\times(b+\beta)(b+\beta-1)-2(a-\alpha)(a+\alpha)(b+\beta)(b-\beta)],$$

$$\begin{pmatrix} a & b & c \\ \frac{1}{2} & -\frac{1}{2} & 0 \end{pmatrix} = (-)^{\frac{1}{2}(a+b+k-2)}\frac{2\Delta(abc)}{[(2a+1)(2b+1)]^{\frac{1}{2}}}\times$$

$$\times\frac{[\frac{1}{2}(k+a+b)]!}{[\frac{1}{2}(a+b-k)]!\ [\frac{1}{2}(a+k-b-1)]!\ [\frac{1}{2}(b+k-a-1)]!}$$

with $k = c$ if $a+b+c$ even, $k = c+1$ if $a+b+c$ odd, and $\Delta(abc)$ as in (2.34).

$$\begin{pmatrix} a & b & c \\ \alpha & -\alpha-c & c \end{pmatrix} = (-)^{a-\alpha-2c}\times$$

$$\times\left[\frac{(2c)!\ (a+b-c)!\ (a-\alpha)!\ (c+b+\alpha)!}{(a+b+c+1)!\ (c-a+b)!\ (c+a-b)!\ (b-c-\alpha)!\ (a+\alpha)!}\right]^{\frac{1}{2}}.$$

By specializing some of the relations in Appendix II:

$$\sum_{c\text{ even}}(2c+1)\begin{pmatrix}a & b & c\\ \frac{1}{2} & -\frac{1}{2} & 0\end{pmatrix}^2=\sum_{c\text{ odd}}(2c+1)\begin{pmatrix}a & b & c\\ \frac{1}{2} & -\frac{1}{2} & 0\end{pmatrix}^2=\tfrac{1}{2}$$

$$\sum_{\beta}\beta(2c+1)\begin{pmatrix}a & b & c\\ \alpha & \beta & -\gamma\end{pmatrix}^2=\langle\beta\rangle_{\text{Av.}}=\gamma\frac{c(c+1)+b(b+1)-a(a+1)}{2c(c+1)},$$

$$\begin{pmatrix}a & b & c\\ 1 & -1 & 0\end{pmatrix}=\begin{pmatrix}a & b & c\\ 0 & 0 & 0\end{pmatrix}\frac{c(c+1)-a(a+1)-b(b+1)}{2[a(a+1)b(b+1)]^{\frac{1}{2}}}$$

if $a+b+c$ even,

$$\begin{pmatrix}a & b & c\\ 1 & 1 & -2\end{pmatrix}=\begin{pmatrix}a & b & c\\ 1 & -1 & 0\end{pmatrix}(b-a)(a+b+1)\left[\frac{(c-2)!}{(c+2)!}\right]^{\frac{1}{2}}$$

if $a+b+c$ odd.

$$\begin{pmatrix}a & a & c\\ 0 & 0 & 0\end{pmatrix}=\begin{pmatrix}a-1 & a+1 & c\\ 0 & 0 & 0\end{pmatrix}\left[\frac{(c+2)(c-1)}{c(c+1)}\right]^{\frac{1}{2}}$$

$$=-\begin{pmatrix}a-1 & a-1 & c\\ 0 & 0 & 0\end{pmatrix}\left[\frac{(2a+c)(2a-c-1)}{(2a-c)(2a+c+1)}\right]^{\frac{1}{2}}$$

$$=\begin{pmatrix}a & a-1 & c\\ 1 & -1 & 0\end{pmatrix}2a\left[\frac{(a+1)(a-1)}{c(c+1)(2a-c)(2a+c+1)}\right]^{\frac{1}{2}}$$

$$\begin{pmatrix}a & a & c\\ 1 & 1 & -2\end{pmatrix}=\begin{pmatrix}a & a & c\\ 0 & 0 & 0\end{pmatrix}\left[\frac{c(c+1)}{(c-1)(c+2)}\right]^{\frac{1}{2}}\quad\text{if } c \text{ is even.}$$

$$\begin{pmatrix}a & a & c\\ a & -a & 0\end{pmatrix}=\begin{pmatrix}a & a & c\\ a-1 & 1-a & 0\end{pmatrix}\frac{2a}{c(c+1)-2a}$$

$$\begin{pmatrix}a & b & c\\ \frac{1}{2} & \frac{1}{2} & -1\end{pmatrix}=-\frac{1}{2}\begin{pmatrix}a & b & c\\ \frac{1}{2} & -\frac{1}{2} & 0\end{pmatrix}\frac{(2b+1)+(-)^{a+b+c}(2a+1)}{[c(c+1)]^{\frac{1}{2}}}$$

$$\begin{pmatrix}a & a & c\\ \frac{1}{2} & -\frac{1}{2} & 0\end{pmatrix}=\begin{pmatrix}a & a+1 & c\\ \frac{1}{2} & -\frac{1}{2} & 0\end{pmatrix}\left[\frac{(2a+3)(2a+c+2)(2a-c+1)}{c(c+1)(2a+1)}\right]^{\frac{1}{2}}$$

$$=-\begin{pmatrix}a & a-1 & c\\ \frac{1}{2} & -\frac{1}{2} & 0\end{pmatrix}\left[\frac{(2a-1)(2a-c)(2a+c+1)}{c(c+1)(2a+1)}\right]^{\frac{1}{2}}$$

$$=-\begin{pmatrix}a & a+2 & c\\ \frac{1}{2} & -\frac{1}{2} & 0\end{pmatrix}\left[\frac{(c-1)(c+2)(2a+5)(2a+c+2)(2a-c+1)}{c(c+1)(2a+1)(2a+c+3)(2a-c+2)}\right]^{\frac{1}{2}}$$

if c is even.

Clebsch–Gordan coefficients, symmetries:

$$\langle ab\alpha\beta|c\gamma\rangle = (-)^{a+b-c}\langle ab-\alpha-\beta|c-\gamma\rangle,$$
$$= (-)^{a+b-c}\langle ba\beta\alpha|c\gamma\rangle,$$
$$= \left(\frac{2c+1}{2b+1}\right)^{\frac{1}{2}}(-)^{a-\alpha}\langle ac\alpha-\gamma|b-\beta\rangle,$$
$$= \left(\frac{2c+1}{2a+1}\right)^{\frac{1}{2}}(-)^{b+\beta}\langle cb-\gamma\beta|a-\alpha\rangle.$$

Special cases:

$$\langle ab\alpha\beta|00\rangle = (-)^{a-\alpha}(2a+1)^{-\frac{1}{2}}\delta_{ab}\delta_{\alpha,-\beta},$$
$$\langle a0\alpha0|c\gamma\rangle = \delta_{ac}\,\delta_{\alpha\gamma}.$$

APPENDIX II

6-j SYMBOLS AND RACAH COEFFICIENTS

THE Racah coefficient is defined by the transformation (3.8),

$$|(ab)e, d; c\rangle = \sum_f |a, (bd)f; c\rangle[(2e+1)(2f+1)]^{\frac{1}{2}}\,W(abcd; ef).$$

Orthogonality:

$$\sum_e (2e+1)(2f+1)W(abcd; ef)W(abcd; eg) = \delta_{fg}.$$

Symmetry: (giving 144 equivalent coefficients [38], [52])

$$W(abcd; ef) = W(badc; ef) = W(cdab; ef) = W(acbd; fe), \text{ etc.}$$
$$= (-)^{b+c-e-f}\,W(aefd; bc), \text{ etc.}$$

also $\quad = W(aBCd; EF)$, where

$$B = \tfrac{1}{2}(b+c+e-f),\ C = \tfrac{1}{2}(b+c+f-e),\ E = \tfrac{1}{2}(b+e+f-c),$$
$$F = \tfrac{1}{2}(c+e+f-b).$$

Sum rules [5], [7], [23], [48]

$$W(abcd; ef) = \sum_g (-)^{b+g-c}(2g+1)W(gabf; dc)W(gdbe; ac),$$

$$W(abcd; ef)W(abgh; ei) = \sum_j (2j+1)W(jgfa; ci)W(jdib; hf)\times$$
$$\times W(jgde; ch) = \sum_{jk} (2j+1)(2k+1)W(cifg; ka)W(cijb; kh)\times$$
$$\times W(gfjb; kd)W(cdhg; ej).$$

Contraction of 3-j symbols:

$$W(abcd;\, ef) = \sum (2c+1)(-)^{f-e-\alpha-\delta}\begin{pmatrix} a & b & e \\ \alpha & \beta & -\epsilon \end{pmatrix}\begin{pmatrix} d & c & e \\ \delta & \gamma & \epsilon \end{pmatrix}\times \begin{pmatrix} b & d & f \\ \beta & \delta & -\phi \end{pmatrix}\begin{pmatrix} c & a & f \\ \gamma & \alpha & \phi \end{pmatrix},$$

summed over all z-components *except* γ,

$$W(abcd;\, ef)\begin{pmatrix} c & a & f \\ \gamma & \alpha & \phi \end{pmatrix} = \sum_{\beta\delta\epsilon} (-)^{f-e-\alpha-\delta}\begin{pmatrix} a & b & e \\ \alpha & \beta & -\epsilon \end{pmatrix}\begin{pmatrix} d & c & e \\ \delta & \gamma & \epsilon \end{pmatrix}\begin{pmatrix} b & d & f \\ \beta & \delta & -\phi \end{pmatrix},$$

$$\sum_{f} (2f+1)\, W(abcd;\, ef)\begin{pmatrix} c & a & f \\ \gamma & \alpha & \phi \end{pmatrix}\begin{pmatrix} b & d & f \\ \beta & \delta & -\phi \end{pmatrix}(-)^{f-e-\alpha-\delta} = \begin{pmatrix} a & b & e \\ \alpha & \beta & -\epsilon \end{pmatrix}\begin{pmatrix} d & c & e \\ \delta & \gamma & \epsilon \end{pmatrix}.$$

$$\sum_{cf} (2c+1)(2f+1)\, W(abcd;\, ef)\begin{pmatrix} c & a & f \\ \gamma & \alpha & \phi \end{pmatrix}\begin{pmatrix} b & d & f \\ \beta & \delta & -\phi \end{pmatrix}\begin{pmatrix} d & c & e \\ \delta & \gamma & \epsilon \end{pmatrix}\times \times(-)^{f-e-\alpha-\delta} = \begin{pmatrix} a & b & e \\ \alpha & \beta & -\epsilon \end{pmatrix},$$

$$\sum_{cef} (2c+1)(2e+1)(2f+1)\, W(abcd;\, ef)\begin{pmatrix} c & a & f \\ \gamma & \alpha & \phi \end{pmatrix}\begin{pmatrix} b & d & f \\ \beta & \delta & -\phi \end{pmatrix}\times \times\begin{pmatrix} d & c & e \\ \delta & \gamma & \epsilon \end{pmatrix}\begin{pmatrix} a & b & e \\ \alpha & \beta & -\epsilon \end{pmatrix}(-)^{f-e-\alpha-\delta} = 1.$$

If $a+b+e$ is even, a special case is

$$\begin{pmatrix} c & d & e \\ -\frac{1}{2} & \frac{1}{2} & 0 \end{pmatrix} = -[(2a+1)(2b+1)]^{\frac{1}{2}}\, W(abcd;\, e\tfrac{1}{2})\begin{pmatrix} a & b & e \\ 0 & 0 & 0 \end{pmatrix}.$$

Algebraic formula for the general Racah coefficient is given by equation (3.15), and for $e = \frac{1}{2}$ and 1, in Table 4. Other special cases are

$$W(abcd;\, a+b,\, f) = \left[\frac{(2a)!\,(2b)!\,(a+b+c+d+1)!\,(a+b+c-d)!}{(2a+2b+1)!\,(c+d-a-b)!\,(a+c-f)!\,(a+f-c)!}\times \times\frac{(a+b+d-c)!\,(c+f-a)!\,(d+f-b)!}{(a+c+f+1)!\,(b+d-f)!\,(b+f-d)!\,(b+f+d+1)!}\right]^{\frac{1}{2}},$$

$$W(abcd, of) = \frac{(-)^{(a+c-f)}}{[(2a+1)(2c+1)]^{\frac{1}{2}}}\,\delta(a, b)\,\delta(c, d).$$

6-*j symbol, definition*:

$$\begin{Bmatrix} a & b & e \\ d & c & f \end{Bmatrix} = (-)^{a+b+c+d}\,W(abcd;ef).$$

Triangular conditions: the four triangular conditions which must be satisfied by the six angular momenta in the 6-*j* symbol may be illustrated in the following way:

$$\begin{Bmatrix} \circ - \circ - \circ \\ \quad \end{Bmatrix},\ \begin{Bmatrix} \circ \searrow \quad \\ \quad \circ - \circ \end{Bmatrix},\ \begin{Bmatrix} \quad \nearrow \circ \searrow \quad \\ \circ \quad\quad \circ \end{Bmatrix},\ \begin{Bmatrix} \quad\quad \nearrow \circ \\ \circ - \circ \quad \end{Bmatrix}$$

Symmetries: the 6-*j* symbol is invariant for interchange of any two columns, and also for interchange of the upper and lower arguments in each of any two columns, i.e.

$$\begin{Bmatrix} a & b & e \\ d & c & f \end{Bmatrix} = \begin{Bmatrix} a & e & b \\ d & f & c \end{Bmatrix} = \begin{Bmatrix} e & b & a \\ f & c & d \end{Bmatrix} = \begin{Bmatrix} a & c & f \\ d & b & e \end{Bmatrix} = \begin{Bmatrix} d & c & e \\ a & b & f \end{Bmatrix}, \text{ etc.}$$

Contraction of 3-j symbols:

$$\sum_{\alpha\beta\gamma\alpha'\beta'} (-)^{A+B+C+\alpha+\beta+\gamma} \begin{pmatrix} A & B & c \\ \alpha & -\beta & \gamma' \end{pmatrix} \begin{pmatrix} B & C & a \\ \beta & -\gamma & \alpha' \end{pmatrix} \begin{pmatrix} C & A & b \\ \gamma & -\alpha & \beta' \end{pmatrix} \begin{pmatrix} a & b & c_1 \\ \alpha' & \beta' & \gamma_1' \end{pmatrix}$$

$$= \frac{1}{2c+1}\,\delta_{cc_1}\delta_{\gamma'\gamma_1'} \begin{Bmatrix} a & b & c \\ A & B & C \end{Bmatrix},$$

$$\sum_{\alpha\beta\gamma} (-)^{A+B+C+\alpha+\beta+\gamma} \begin{pmatrix} A & B & c \\ \alpha & -\beta & \gamma' \end{pmatrix} \begin{pmatrix} B & C & a \\ \beta & -\gamma & \alpha' \end{pmatrix} \begin{pmatrix} C & A & b \\ \gamma & -\alpha & \beta' \end{pmatrix}$$

$$= \begin{pmatrix} a & b & c \\ \alpha' & \beta' & \gamma' \end{pmatrix} \begin{pmatrix} a & b & c \\ A & B & C \end{pmatrix},$$

Sum rules:

$$\sum_{k} (-1)^{2k}(2k+1) \begin{Bmatrix} a & b & k \\ a & b & f \end{Bmatrix} = 1,$$

$$\sum_{k} (-)^{a+b+k}(2k+1) \begin{Bmatrix} a & b & k \\ b & a & f \end{Bmatrix} = \delta_{f0}\{(2a+1)(2b+1)\}^{\frac{1}{2}},$$

$$\sum_{k} (2k+1)(2f+1)\begin{Bmatrix} a & b & k \\ c & d & f \end{Bmatrix}\begin{Bmatrix} a & b & k \\ c & d & g \end{Bmatrix} = \delta_{fg},$$

$$\sum_{k} (-)^{f+g+k}(2k+1)\begin{Bmatrix} a & b & k \\ c & d & f \end{Bmatrix}\begin{Bmatrix} a & b & k \\ d & c & g \end{Bmatrix} = \begin{Bmatrix} a & d & f \\ b & c & g \end{Bmatrix}.$$

APPENDIX III

9-j SYMBOLS OR X-COEFFICIENTS

THE X of Fano is defined by the transformation (3.23), with

$$\frac{\langle (ab)c, (de)f; i | (ad)g, (be)h; i \rangle}{[(2c+1)(2f+1)(2g+1)(2h+1)]^{\frac{1}{2}}} = \begin{Bmatrix} a & b & c \\ d & e & f \\ g & h & i \end{Bmatrix}$$

$$\equiv X(abc, def, ghi).$$

Orthogonality:

$$\sum_{cf} (2c+1)(2f+1)(2g+1)(2h+1)\begin{Bmatrix} a & b & c \\ d & e & f \\ g & h & i \end{Bmatrix}\begin{Bmatrix} a & b & c \\ d & e & f \\ j & k & i \end{Bmatrix} = \delta_{gj}\,\delta_{hk}.$$

Symmetry: (72 relations [39]). The X is invariant under interchange of rows and columns (reflection about a diagonal) and is multiplied by $(-)^p$

$$(\text{where } p = a+b+c+d+e+f+g+h+i)$$

upon interchange of two adjacent rows or columns.

Sum rule: (others are given in the literature [2], [61])

$$\sum_{jk}(-)^{2e+k-f-h}(2j+1)(2k+1)\begin{Bmatrix} a & b & c \\ e & d & f \\ j & k & i \end{Bmatrix}\begin{Bmatrix} a & e & j \\ d & b & k \\ g & h & i \end{Bmatrix} = \begin{Bmatrix} a & b & c \\ d & e & f \\ g & h & i \end{Bmatrix}$$

Contraction of Racah coefficients:

$$X(abc, def, ghi) = \sum_{k} (2k+1)\, W(aidh; kg)\, W(bfhd; ke)\, W(aibf; kc),$$

$$\sum_{c} (2c+1)\, W(aibf; kc) X(abc, def, ghi) = W(aidh; kg)\, W(bfhd; ke),$$

etc.

$$W(ghjk;\, il)X(abc,\, def,\, ghi) = \sum_{st} (2s+1)(2t+1)\; W(cfjk;\, is)\times$$
$$\times W(desk;\, ft)W(belk;\, ht)X(abc,\, dts,\, glj).$$

Contraction of 3-j symbols:

$$\begin{Bmatrix} a & b & c \\ d & e & f \\ g & h & i \end{Bmatrix} = (2a+1)\sum \begin{pmatrix} a & b & c \\ \alpha & \beta & \gamma \end{pmatrix}\begin{pmatrix} b & e & h \\ \beta & \epsilon & \eta \end{pmatrix}\begin{pmatrix} c & f & i \\ \gamma & \phi & \nu \end{pmatrix}\begin{pmatrix} a & d & g \\ \alpha & \delta & \rho \end{pmatrix}\times$$
$$\times\begin{pmatrix} d & e & f \\ \delta & \epsilon & \phi \end{pmatrix}\begin{pmatrix} g & h & i \\ \rho & \eta & \nu \end{pmatrix},$$

summed over all z-components *except* α.

$$\begin{pmatrix} a & b & c \\ \alpha & \beta & \gamma \end{pmatrix}\begin{Bmatrix} a & b & c \\ d & e & f \\ g & h & i \end{Bmatrix} = \sum_{\epsilon\eta\phi\nu\delta\rho} \begin{pmatrix} b & e & h \\ \beta & \epsilon & \eta \end{pmatrix}\begin{pmatrix} c & f & i \\ \gamma & \phi & \nu \end{pmatrix}\begin{pmatrix} a & d & g \\ \alpha & \delta & \rho \end{pmatrix}\begin{pmatrix} d & e & f \\ \delta & \epsilon & \phi \end{pmatrix}\times$$
$$\begin{pmatrix} g & h & i \\ \rho & \eta & \nu \end{pmatrix},\qquad \sum_b (2b+1)\begin{pmatrix} a & b & c \\ \alpha & \beta & \gamma \end{pmatrix}\begin{pmatrix} b & e & h \\ \beta & \epsilon & \eta \end{pmatrix}\begin{Bmatrix} a & b & c \\ d & e & f \\ g & h & i \end{Bmatrix}$$
$$= \sum_{\phi\nu\delta\rho} \begin{pmatrix} c & f & i \\ \gamma & \phi & \nu \end{pmatrix}\begin{pmatrix} a & d & g \\ \alpha & \delta & \rho \end{pmatrix}\begin{pmatrix} d & e & f \\ \delta & \epsilon & \phi \end{pmatrix}\begin{pmatrix} g & h & i \\ \rho & \eta & \nu \end{pmatrix},$$

$$\sum_{bc} (2b+1)(2c+1)\begin{pmatrix} a & b & c \\ \alpha & \beta & \gamma \end{pmatrix}\begin{pmatrix} b & e & h \\ \beta & \epsilon & \eta \end{pmatrix}\begin{pmatrix} c & f & i \\ \gamma & \phi & \nu \end{pmatrix}\begin{Bmatrix} a & b & c \\ d & e & f \\ g & h & i \end{Bmatrix}$$
$$= \begin{pmatrix} a & d & g \\ \alpha & \delta & \rho \end{pmatrix}\begin{pmatrix} d & e & f \\ \delta & \epsilon & \phi \end{pmatrix}\begin{pmatrix} g & h & i \\ \rho & \eta & \nu \end{pmatrix}, \text{ etc.}$$

Special cases:

$$\begin{Bmatrix} a & b & c \\ d & e & f \\ g & h & 0 \end{Bmatrix} = \frac{\delta_{cf}\,\delta_{gh}(-)^{c+g-a-e}\, W(abde;\, cg)}{[(2c+1)(2g+1)]^{\frac{1}{2}}},$$

$$\begin{Bmatrix} a & b & c \\ d & e & c \\ g & g & 1 \end{Bmatrix} = \frac{a(a+1)-d(d+1)-b(b+1)+e(e+1)}{[4c(c+1)(2c+1)g(g+1)(2g+1)]^{\frac{1}{2}}}\times$$
$$\times(-)^{c+g-a-e}\, W(abde;\, cg).$$

With $g = \frac{1}{2}$, $c+d+e$ even, and using A.1, this gives

$$[6(2c+1)(2d+1)(2e+1)]^{\frac{1}{2}}\begin{pmatrix} c & d & e \\ 0 & 0 & 0 \end{pmatrix}\begin{Bmatrix} a & b & c \\ d & e & c \\ \frac{1}{2} & \frac{1}{2} & 1 \end{Bmatrix} = \begin{pmatrix} a & b & c \\ \frac{1}{2} & \frac{1}{2} & -1 \end{pmatrix}.$$

When $c+d+e$ is odd, we have two other relations

$$\begin{pmatrix} c+1 & d & e \\ 0 & 0 & 0 \end{pmatrix}\begin{Bmatrix} a & b & c \\ d & e & c+1 \\ \frac{1}{2} & \frac{1}{2} & 1 \end{Bmatrix}$$

$$= \frac{(-)^{b+e+\frac{1}{2}}[(d-a)(2a+1)+(e-b)(2b+1)+c+1]}{[6(c+1)(2c+1)(2c+3)(2d+1)(2e+1)]^{\frac{1}{2}}} \times \begin{pmatrix} a & b & c \\ \frac{1}{2} & -\frac{1}{2} & 0 \end{pmatrix},$$

$$\begin{pmatrix} c-1 & d & e \\ 0 & 0 & 0 \end{pmatrix}\begin{Bmatrix} a & b & c \\ d & e & c-1 \\ \frac{1}{2} & \frac{1}{2} & 1 \end{Bmatrix}$$

$$= \frac{(-)^{b+e+\frac{1}{2}}[(d-a)(2a+1)+(e-b)(2b+1)-c]}{[6c(2c+1)(2c-1)(2d+1)(2e+1)]^{\frac{1}{2}}} \begin{pmatrix} a & b & c \\ \frac{1}{2} & -\frac{1}{2} & 0 \end{pmatrix}.$$

Algebraic formulae for $g = h = \frac{1}{2}$ are easily obtained from these.

APPENDIX IV

SPHERICAL HARMONICS

THESE are defined with the same phase as Condon and Shortley [17],

$$Y_{kq} = (2k+1/4\pi)^{\frac{1}{2}} C_{kq}, \text{ where}$$

$$C_{kq}(\theta\phi) = (-)^q \left[\frac{(k-q)!}{(k+q)!}\right]^{\frac{1}{2}} \quad P_k^q(\theta)e^{iq\phi}, \quad \text{if} \quad q \geqslant 0,$$

and

$$C_{k-q}(\theta\phi) = (-)^q C_{kq}(\theta\phi)^*.$$

The P_k^q are the associated Legendre polynomials, with $P_k^0 = P_k$; their properties are well known [40], [42]. Special cases are

$$C_{00} = 1; \qquad C_{10} = \cos\theta; \qquad C_{1\pm 1} = \mp(\tfrac{1}{2})^{\frac{1}{2}} \sin\theta\, e^{\pm i\phi};$$

$$C_{20} = \tfrac{1}{2}(3\cos^2\theta - 1); \qquad C_{2\pm 1} = \mp(\tfrac{3}{2})^{\frac{1}{2}} \cos\theta \sin\theta\, e^{\pm i\phi};$$

$$C_{2\pm 2} = (\tfrac{3}{8})^{\frac{1}{2}} \sin^2\theta\, e^{\pm 2i\phi}$$

Orthogonality:

$$(2k+1)\int C_{kq}(\theta\phi)^* C_{KQ}(\theta\phi)\sin\theta\, d\theta d\phi = \delta_{kK}\,\delta_{qQ}4\pi.$$

Sum rules:

$$\sum_q |C_{kq}(\theta\phi)|^2 = 1,$$

$$\sum_k (2k+1)C_{k0}(\theta\phi) = 2\delta(\cos\theta - 1).$$

Addition theorems: $\sum_q C_{kq}(\theta\phi)C_{kq}(\theta'\phi')^* = P_k(\cos\omega)$

if ω is the angle between the two directions $(\theta\phi)$ and $(\theta'\phi')$.

$$C_{a\alpha}(\theta\phi)C_{b\beta}(\theta\phi) = \sum_c C_{c\gamma}(\theta\phi)(2c+1)(-)^{\gamma}\begin{pmatrix} a & b & c \\ \alpha & \beta & -\gamma \end{pmatrix}\begin{pmatrix} a & b & c \\ 0 & 0 & 0 \end{pmatrix},$$

$$\sum_{\alpha\beta} C_{a\alpha}(\theta\phi)C_{b\beta}(\theta\phi)\begin{pmatrix} a & b & c \\ \alpha & \beta & -\gamma \end{pmatrix} = C_{c\gamma}(\theta\phi)(-)^{\gamma}\begin{pmatrix} a & b & c \\ 0 & 0 & 0 \end{pmatrix},$$

$$\int C_{a\alpha}(\theta\phi)C_{b\beta}(\theta\phi)C_{c\gamma}(\theta\phi)\sin\theta\, d\theta d\phi = 4\pi\begin{pmatrix} a & b & c \\ \alpha & \beta & \gamma \end{pmatrix}\begin{pmatrix} a & b & c \\ 0 & 0 & 0 \end{pmatrix},$$

$$\therefore \int P_a(\cos\theta)\; P_b(\cos\theta)\; P_c(\cos\theta)\sin\theta\, d\theta = 2\begin{pmatrix} a & b & c \\ 0 & 0 & 0 \end{pmatrix}^2.$$

APPENDIX V

ROTATION MATRIX ELEMENTS

OUR definition of $\mathscr{D}^j_{mm'}(\alpha\beta\gamma)$ for the rotation of axes through $(\alpha\beta\gamma)$ is the same as that of Rose [54], Messiah [46]. Bohr and Mottelson [13] use rotation matrices which are the complex conjugate of ours. Wigner [78], Fano and Racah [31], Edmonds [22] and Rose [53] use the same notation, but $(\alpha\beta\gamma)$ are then Euler angles for a rotation of the system. Associated with this is the equivalent coordinate rotation $(\gamma\,\beta\,\alpha)^{-1} = (-\alpha\ -\beta\ -\gamma)$, so their usage is equivalent to a different sign convention for the angles of rotation.

With our definition (equation (2.17))

$$\mathscr{D}^j_{mm'}(\alpha\beta\gamma) = e^{-im\alpha}\, d^j_{mm'}(\beta)e^{-im'\gamma}.$$

The formula for $d^j_{mm'}(\beta)$ is given by equation (2.18), and given explicitly for $j = \frac{1}{2}$, 1, $\frac{3}{2}$ and 2 in Table 1.

Orthogonality:

$$(2j+1)\int \mathscr{D}^{j}_{mm'}{}^* \mathscr{D}^{J}_{MM'} \sin\beta d\beta d\alpha d\gamma = \delta_{jJ}\,\delta_{mM}\,\delta_{m'M'}\,8\pi^2.$$

Sum rule:

$$\sum_{m'} \mathscr{D}^{j}_{mm'}(\mathscr{D}^{j}_{m''m'})^* = \delta_{mm''}.$$

Closure:

$$\sum_{m''} \mathscr{D}^{j}_{mm''}(\alpha_2\beta_2\gamma_2)\mathscr{D}^{j}_{m''m'}(\alpha_1\beta_1\gamma_1) = \mathscr{D}^{j}_{mm'}(\alpha\beta\gamma),$$

where $(\alpha\beta\gamma)$ is the resultant of first $(\alpha_1\beta_1\gamma_1)$ then $(\alpha_2\beta_2\gamma_2)$,

$$\therefore \sum_{m''} d^{j}_{mm''}(\beta_2)d^{j}_{m''m'}(\beta_1) = d^{j}_{mm'}(\beta_1+\beta_2).$$

Symmetry:

$$d^{j}_{mm'}(\beta) = (-)^{m-m'}d^{j}_{m'm}(\beta) = d^{j}_{-m'-m}(\beta) = d^{j}_{m'm}(-\beta)$$
$$= (-)^{j-m}d^{j}_{m-m'}(\pi-\beta) = (-)^{j+m'}d^{j}_{m-m'}(\pi+\beta),$$

$$\therefore \mathscr{D}^{j}_{mm'}(\alpha\beta\gamma)^* = (-)^{m-m'}\mathscr{D}^{j}_{-m-m'}(\alpha\beta\gamma) = \mathscr{D}^{j}_{m'm}(-\gamma-\beta-\alpha),$$

where $(-\gamma\ -\beta\ -\alpha)$ is the rotation inverse to $(\alpha\beta\gamma)$.

Special cases:

$$\mathscr{D}^{j}_{m0}(\alpha\beta\gamma) = C_{jm}(\beta\alpha)^*;$$

$$\therefore d^{j}_{m0}(\beta) = (-)^m\left[\frac{(j-m)!}{(j+m)!}\right]^{\frac{1}{2}} P^m_j(\beta) \quad \text{if} \quad m \geqslant 0,$$

and

$$d^{j}_{00}(\beta) = P_j(\cos\beta).$$

$$d^{j}_{jm}(\beta) = (-1)^{j-m}[(2j)!/(j+m)!\,(j-m)!]^{\frac{1}{2}}\times$$
$$\times(\cos\tfrac{1}{2}\beta)^{j+m}(\sin\tfrac{1}{2}\beta)^{j-m}.$$

Contraction:

$$\mathscr{D}^{A}_{aa'}(\alpha\beta\gamma)\mathscr{D}^{B}_{bb'}(\alpha\beta\gamma) = \sum_C (2C+1)\begin{pmatrix}A & B & C\\ a & b & c\end{pmatrix}\begin{pmatrix}A & B & C\\ a' & b' & c'\end{pmatrix}\mathscr{D}^{C}_{cc'}(\alpha\beta\gamma)^*,$$

$$\mathscr{D}^{C}_{cc'}(\alpha\beta\gamma)^*\begin{pmatrix}A & B & C\\ a' & b' & c'\end{pmatrix} = \sum_{ab}\begin{pmatrix}A & B & C\\ a & b & c\end{pmatrix}\mathscr{D}^{A}_{aa'}(\alpha\beta\gamma)\mathscr{D}^{B}_{bb'}(\alpha\beta\gamma),$$

$$\mathscr{D}^C_{cc'}(\alpha\beta\gamma)^* = \sum_{aba'b'} (2C+1)\begin{pmatrix} A & B & C \\ a & b & c \end{pmatrix} \mathscr{D}^A_{aa'}(\alpha\beta\gamma)\mathscr{D}^B_{bb'}(\alpha\beta\gamma)\begin{pmatrix} A & B & C \\ a' & b' & c' \end{pmatrix},$$

$$\sum_{abc} \mathscr{D}^A_{aa'}(\alpha\beta\gamma)\mathscr{D}^B_{bb'}(\alpha\beta\gamma)\mathscr{D}^C_{cc'}(\alpha\beta\gamma)\begin{pmatrix} A & B & C \\ a & b & c \end{pmatrix} = \begin{pmatrix} A & B & C \\ a' & b' & c' \end{pmatrix},$$

$$\int \mathscr{D}^C_{cc'}(\alpha\beta\gamma)\mathscr{D}^A_{aa'}(\alpha\beta\gamma)\mathscr{D}^B_{bb'}(\alpha\beta\gamma) \sin\beta d\beta d\alpha d\gamma$$

$$= 8\pi^2\begin{pmatrix} A & B & C \\ a & b & c \end{pmatrix}\begin{pmatrix} A & B & C \\ a' & b' & c' \end{pmatrix}.$$

APPENDIX VI

TENSORS AND THEIR MATRIX ELEMENTS

THE commutation rules for spherical tensor components with the spherical components $J_\mu (\mu = 0, \pm 1)$ of $\mathbf{J}$ are

$$[J_\mu, T_{kq}] = T_{kq+\mu}[k(k+1)(2k+1)]^{\frac{1}{2}}(-)^{k+q+\mu-1}\begin{pmatrix} k & k & 1 \\ -q-\mu & q & \mu \end{pmatrix}.$$

Basic tensors are the spherical harmonics $\mathbf{C}_k$. When $k = 1$ we have vectors, $\mathbf{a} = a\mathbf{C}_1(\theta\phi) = \sum_\mu (-)^\mu a_\mu \mathbf{e}_{-\mu}$. $(\theta\phi)$ are the polar angles of $\mathbf{a}$, so $\mathbf{C}_1$ is a unit vector along $\mathbf{a}$. The $\mathbf{e}_\mu$ are unit spherical vectors, $\mathbf{e}_0 = \mathbf{e}_z$, $\mathbf{e}_{\pm 1} = \mp(\mathbf{e}_x \pm i\mathbf{e}_y)/2^{\frac{1}{2}}$, so the vector components $a_\mu = aC_{1\mu}(\theta\phi)$. Product tensors are defined by (4.6),

$$T_{KQ}(\mathbf{R}_{k_1}, \mathbf{S}_{k_2}) \equiv T_{KQ}(k_1, k_2)$$

$$= (2K+1)^{\frac{1}{2}}(-)^{2k_2+K-Q}\sum_{q_1q_2}\begin{pmatrix} K & k_1 & k_2 \\ Q & -q_1 & -q_2 \end{pmatrix} R_{k_1q_1}S_{k_2q_2}.$$

When $k_1 = k_2 = 1$, $\sqrt{2}\mathbf{T}_1(\mathbf{a}, \mathbf{b}) = i\mathbf{a} \wedge \mathbf{b}$, and $\mathbf{T}_2(\mathbf{a}, \mathbf{b})$ is given in section 4.5. Other examples are the bipolar harmonics of section 4.6, and the spherical harmonic addition theorem. We may derive various re-coupling relations for such tensors; for example, if tensors k_2 and k_3 commute

$$\mathbf{T}_K(k_1, k_2) \,.\, \mathbf{T}_K(k_3, k_4)$$

$$= (2K+1)(-)^{k_1+k_4}\sum_{K'} W(k_1k_2k_3k_4; KK')\mathbf{T}_{K'}(k_1k_3) \,.\, \mathbf{T}_{K'}(k_2k_4).$$

For example, with $K = 0$, $k_1 = k_2 = k_3 = k_4 = 1$,

$$(\mathbf{a}\,.\,\mathbf{c})(\mathbf{b}\,.\,\mathbf{d}) = \tfrac{1}{3}(\mathbf{a}\,.\,\mathbf{b})(\mathbf{c}\,.\,\mathbf{d})+\tfrac{1}{2}(\mathbf{a}\wedge\mathbf{b})\,.\,(\mathbf{c}\wedge\mathbf{d})+ \\ +\mathbf{T}_2(\mathbf{a},\mathbf{b})\,.\,\mathbf{T}_2(\mathbf{c},\mathbf{d}).$$

When **c**, **d** are Pauli matrices, with $\boldsymbol{\sigma}\,.\,\boldsymbol{\sigma} = 3$, $T_2(\boldsymbol{\sigma},\boldsymbol{\sigma}) = 0$, [20],

$$(\mathbf{a}\,.\,\boldsymbol{\sigma})(\mathbf{b}\,.\,\boldsymbol{\sigma}) = (\mathbf{a}\,.\,\mathbf{b})+i\boldsymbol{\sigma}\,.\,(\mathbf{a}\wedge\mathbf{b}),$$

and when $\mathbf{a} = \mathbf{b} = \mathbf{J}_1$, $\mathbf{c} = \mathbf{d} = \mathbf{J}_2$, we get

$$\mathbf{T}_2(\mathbf{J}_1\mathbf{J}_1)\,.\,\mathbf{T}_2(\mathbf{J}_2\mathbf{J}_2) = (\mathbf{J}_1\,.\,\mathbf{J}_2)^2+\tfrac{1}{2}(\mathbf{J}_1\,.\,\mathbf{J}_2)-\tfrac{1}{3}\mathbf{J}_1^2\mathbf{J}_2^2.$$

Again, if $\mathbf{a} = \mathbf{s}_1$, $\mathbf{b} = \mathbf{s}_2$, $\mathbf{c} = \mathbf{d} = \mathbf{r}$, we get the tensor force S_{12}

$$\mathbf{T}_2(\mathbf{s}_1\mathbf{s}_2)\,.\,\mathbf{T}_2(\mathbf{r}\mathbf{r}) = \mathbf{r}^2S_{12} = (\mathbf{s}_1\,.\,\mathbf{r})(\mathbf{s}_2\,.\,\mathbf{r})-\tfrac{1}{3}(\mathbf{s}_1\,.\,\mathbf{s}_2)\mathbf{r}^2$$

where $T_2(\boldsymbol{rr}) = \sqrt{\left(\frac{2}{3}\right)}r^2C_2(\theta\phi)$. This may be recoupled, using

$$\mathbf{r}(\mathbf{s}\,.\,\mathbf{r})-\tfrac{1}{3}r^2\mathbf{s} = -(10/9)^{\frac{1}{2}}r^2\mathbf{T}_1(\mathbf{C}_2,\mathbf{s})$$

to give $\mathbf{T}_2(\mathbf{s}_1\mathbf{s}_2).\mathbf{T}_2(\mathbf{rr}) = -(10/9)^{\frac{1}{2}}r^2\mathbf{s}_1.\mathbf{T}_1(\mathbf{C}_2,\mathbf{s}_2)$.

Tensors may be formed by 'polarizing' solid harmonics r^kC_{kq} with $n \leqslant k$ vectors **A**, **B**, $\cdots$ [27], [75],

$$T_{kq}(\mathbf{A},\mathbf{B},\cdots\mathbf{H},\mathbf{r}^{k-n}) = \sum_{\lambda\mu\cdots}(-)^{\lambda+\mu+\cdots}\times \\ \times A_\lambda B_\mu\cdots H_\eta\nabla_{-\lambda}\nabla_{-\mu}\cdots\nabla_{-\eta}r^kC_{kq}(\theta\phi),$$

and if **B**, $\cdots$ **H** commute with $\boldsymbol{\nabla}$,

$$= (\mathbf{A}\,.\,\boldsymbol{\nabla})(\mathbf{B}\,.\,\boldsymbol{\nabla})\cdots(\mathbf{H}\,.\,\boldsymbol{\nabla})r^kC_{kq}(\theta\phi).$$

Each step replaces a vector **r** by a vector **A**, **B** $\cdots$, leaving unchanged the transformation properties. Other polarized harmonics may be formed with the operators **L** and $\boldsymbol{\nabla}$; vector harmonics were introduced in section 4.10.2 and are further discussed by Hill [35] and Edmonds [22]. In the product tensor notation above, the vector harmonic is

$$\mathbf{Y}^Q_{Kk1} = T_{KQ}(\mathbf{Y}_k,\mathbf{e}) = \sum_{q\mu}(2K+1)^{\frac{1}{2}}(-)^{K-Q}\begin{pmatrix}K & k & 1\\ Q & -q & -\mu\end{pmatrix}\times \\ \times Y_{kq}(\theta\phi)\mathbf{e}_\mu.$$

Some useful properties of the various tensors follow from

$$L_\mu C_{kq}(\theta\phi) = [k(k+1)(2k+1)]^{\frac{1}{2}}(-)^{k+q+\mu}\begin{pmatrix} k & k & 1 \\ q+\mu & -q & -\mu \end{pmatrix} C_{kq+\mu}(\theta\phi),$$

$$\nabla_\mu[f(r)C_{kq}(\theta\phi)] = -\left[\frac{(k+1)(2k+3)}{2k+1}\right]^{\frac{1}{2}}(-)^{k+q+\mu}\begin{pmatrix} k+1 & k & 1 \\ q+\mu & -q & -\mu \end{pmatrix}\times$$
$$\times C_{k+1,q+\mu}(\theta\phi)\left(\frac{d}{dr}-\frac{k}{r}\right)f(r)+\left[\frac{k(2k-1)}{2k+1}\right]^{\frac{1}{2}}(-)^{k+q+\mu}\times$$
$$\times\begin{pmatrix} k-1 & k & 1 \\ q+\mu & -q & -\mu \end{pmatrix} C_{k-1,q+\mu}(\theta\phi)\left(\frac{d}{dr}+\frac{k+1}{r}\right)f(r),$$

$$\nabla_\mu(r^k C_{kq}) = (-)^{k+q+\mu}[k(2k-1)(2k+1)]^{\frac{1}{2}}\times$$
$$\times\begin{pmatrix} k-1 & k & 1 \\ q+\mu & -q & -\mu \end{pmatrix} r^{k-1}C_{k-1,q+\mu},$$

$$\nabla_\mu(r^{-k-1}C_{kq}) = (-)^{k+q+\mu}[(k+1)(2k+1)(2k+3)]^{\frac{1}{2}}\times$$
$$\times\begin{pmatrix} k+1 & k & 1 \\ q+\mu & -q & -\mu \end{pmatrix} r^{-k-2}C_{k+1,q+\mu}.$$

So we have

$$[k(k+1)]^{\frac{1}{2}}\mathbf{Y}^q_{kk1} = \mathbf{L}(Y_{kq}),$$
$$[k(2k+1)]^{\frac{1}{2}}r^{k-1}\mathbf{Y}^q_{kk-11} = \boldsymbol{\nabla}(r^k Y_{kq}),$$
$$[(k+1)(2k+1)]^{\frac{1}{2}}r^{-k-2}\mathbf{Y}^q_{kk+11} = \boldsymbol{\nabla}(r^{-k-1} Y_{kq}).$$

Further, using the relation

$$\boldsymbol{\nabla}\wedge\mathbf{L}\phi = -i\left[\mathbf{r}\nabla^2\phi-\boldsymbol{\nabla}\left(\frac{\partial}{\partial r}r\phi\right)\right],$$

we get

$$(2k+1)^{\frac{1}{2}}\boldsymbol{\nabla}\wedge\mathbf{L}(fY_{kq}) = ik(k+1)^{\frac{1}{2}}\,\mathbf{Y}^q_{k\,k+11}\left(\frac{d}{dr}-\frac{k}{r}\right)f(r)+$$
$$+i(k+1)k^{\frac{1}{2}}\,\mathbf{Y}^q_{k\,k-11}\left(\frac{d}{dr}+\frac{k+1}{r}\right)f(r),$$

and in particular

$$\boldsymbol{\nabla}\wedge\mathbf{L}(r^kC_{kq}) = i(k+1)\boldsymbol{\nabla}(r^kC_{kq}).$$

Similar tensors arising in radiation theory are discussed in section 6.1.1, and others have been applied to β-decay theory

[3], [45], [57]. We also have

$$(2k+1)\nabla(fC_{kq})\cdot\nabla = [k(2k-1)]^{\frac{1}{2}}\left(\frac{d}{dr}+\frac{k+1}{r}\right)f(r)\,T_{kq}(\mathbf{C}_{k-1},\nabla)$$

$$-[(k+1)(2k+3)]^{\frac{1}{2}}\left(\frac{d}{dr}-\frac{k}{r}\right)f(r)T_{kq}(\mathbf{C}_{k+1},\nabla).$$

Some of the multipole tensor expansions useful in physics are

$$e^{i\mathbf{k}.\mathbf{r}} = \sum_l (2l+1)i^l j_l(kr)\mathbf{C}_l(\theta_k\phi_k)\,.\,\mathbf{C}_l(\theta_r\phi_r)$$

$$\delta(\mathbf{a}-\mathbf{b}) = (4\pi a^2)^{-1}\delta(a-b)\sum_l (2l+1)\mathbf{C}_l(\theta_a\phi_a)\,.\,\mathbf{C}_l(\theta_b\phi_b),$$

$$e^{-\gamma(\mathbf{a}-\mathbf{b})^2} = \sum_l i^{-l}(2l+1)e^{-\gamma(a^2+b^2)}j_l(2i\gamma ab)\mathbf{C}_l(\theta_a\phi_a)\,.\,\mathbf{C}_l(\theta_b\phi_b);$$

and if $\boldsymbol{\rho} = \mathbf{b}-\mathbf{a}$, with $b > a$,

$$1/\rho = \sum_l (a^l/b^{l+1})\mathbf{C}_l(\theta_a\phi_a)\,.\,\mathbf{C}_l(\theta_b\phi_b)$$

$$e^{ik\rho}/\rho = ikh_0^{(1)}(k\rho) = ik\sum_l (2l+1)j_l(ka)h_l^{(1)}(kb)\mathbf{C}_l(\theta_a\phi_a)\,.\,\mathbf{C}_l(\theta_b\phi_b),$$

$$e^{-\alpha\rho}/\alpha\rho = -\sum_l (2l+1)j_l(i\alpha a)h_l^{(1)}(i\alpha b)\,\mathbf{C}_l(\theta_a\phi_a)\,.\,\mathbf{C}_l(\theta_b\phi_b);$$

if $\mathbf{r} = \mathbf{a}+\mathbf{b}$

$$r^l C_{lm} = \sum_{\lambda\mu}\left(\frac{2l!}{2\lambda!2(l-\lambda)!}\right)^{\frac{1}{2}} a^{l-\lambda}b^{\lambda}\times$$

$$\times\, C_{l-\lambda,m-\mu}(\theta_a\phi_a)C_{\lambda\mu}(\theta_b\phi_b)\,\langle l-\lambda\lambda m-\mu\mu|lm\rangle.$$

Reduced matrix elements, definition:

The Wigner–Eckart theorem (4.15) states

$$\langle JM|T_{kq}|J'M'\rangle = (-)^{2k}\langle JM|J'kM'q\rangle\langle J\|\mathbf{T}_k\|J'\rangle$$

$$= (-)^{J-M}\begin{pmatrix} J & k & J' \\ -M & q & M' \end{pmatrix}(2J+1)^{\frac{1}{2}}\langle J\|\mathbf{T}_k\|J'\rangle.$$

The reduced matrix element $(J\|\mathbf{T}_k\|J')$ defined and used by Racah [31, 48] and Edmonds [22] is related to ours by the equation

$$(J\|T_k\|J') = (2J+1)^{\frac{1}{2}}\langle J\|\mathbf{T}_k\|J'\rangle.$$

The factor $(-)^{2k}$ is included in our definition so that the phases of both reduced matrix elements should be the same. It is relevant only if k is half-integral (for example if T_{kq} is a creation operator for a spin-$\frac{1}{2}$ particle).

Reduction formulae in terms of 6-j symbols:

For convenience we collect the principle reduction formulae of section 5.3 and rewrite them in terms of 6-j symbols. In a two-component system the tensor $\mathbf{R}_{k_1}$ (1) acts only on the first part and $\mathbf{S}_{k_2}(2)$ only on the second part. If

$$T_{KQ}(k_1k_2) = \sum_{q_1q_2} R_{k_1q_1}(1)\, S_{k_2q_2}(2)\ \langle k_1k_2q_1q_2|KQ\rangle$$

then formula (5.12) gives

$$\langle j_1j_2J\|\mathbf{T}_K(k_1k_2)\|j_1'j_2'J'\rangle = \{(2J'+1)(2K+1)\}^{\frac{1}{2}}\begin{Bmatrix} J & J' & K \\ j_1 & j_1' & k_1 \\ j_2 & j_2' & k_2 \end{Bmatrix} \times$$

$$\times\ (2j_1+1)^{\frac{1}{2}}\langle j_1\|\mathbf{R}_{k_1}\|j_1'\rangle(2j_2+1)^{\frac{1}{2}}\langle j_2\|\mathbf{S}_{k_2}\|j_2'\rangle.$$

Special cases of this result are

$\mathbf{S} = 1 \quad , \quad K = k_1 = k$ (equation (5.9)),

$$\langle j_1j_2J\|\mathbf{R}_k(1)\|j_1'j_2'J'\rangle = \delta(j_2j_2')(2J'+1)^{\frac{1}{2}}\begin{Bmatrix} J & J' & k \\ j_1' & j_1 & j_2 \end{Bmatrix} \times$$

$$\times\ (-)^{k+j_2+J'+j_1}(2j_1+1)^{\frac{1}{2}}\langle j_1\|\mathbf{R}_k(1)\|j_1'\rangle;$$

$\mathbf{R} = 1 \quad , \quad K = k_2 = k$

$$\langle j_1j_2J\|S_k(2)\|j_1'j_2'J'\rangle = \delta(j_1j_1')(2J'+1)^{\frac{1}{2}}\begin{Bmatrix} J & J' & k \\ j_2' & j_2 & j_1 \end{Bmatrix} \times$$

$$\times\ (-)^{k+j_1+J+j_2'}(2j_2+1)^{\frac{1}{2}}\langle j_2\|\mathbf{S}_k(2)\|j_2'\rangle;$$

$K = 0 \quad , \quad k_1 = k_2 = k$ (equation (5.13)),

$$\mathbf{R}_k.\mathbf{S}_k = (-)^k(2k+1)^{\frac{1}{2}}T_{00}(kk)$$

$$\langle j_1j_2J\|\mathbf{R}_k\,.\,\mathbf{S}_k\|j_1'j_2'J'\rangle = \delta(JJ')(-)^{j_1'+j_2+J}\begin{Bmatrix} j_1 & j_1' & k \\ j_2' & j_2 & J \end{Bmatrix} \times$$

$$\times\ (2j_1+1)^{\frac{1}{2}}\langle j_1\|\mathbf{R}_k\|j_1'\rangle(2j_2+1)^{\frac{1}{2}}\langle j_2\|\mathbf{S}_k\|j_2'\rangle.$$

Basic reduced matrices are

$$\langle J\|\mathbf{J}\|J'\rangle = [J(J+1)]^{\frac{1}{2}}\delta_{JJ'},$$

$$\langle l\|\mathbf{C}_k\|l'\rangle = (2l'+1)^{\frac{1}{2}}(-)^{l}\begin{pmatrix} l & k & l' \\ 0 & 0 & 0 \end{pmatrix},$$

$$\langle l\|\nabla\|l'\rangle = (2l'+1)^{\frac{1}{2}}(-)^{l}\begin{pmatrix} l & 1 & l' \\ 0 & 0 & 0 \end{pmatrix}\left\langle l\left|\frac{d}{dr}+\frac{a}{r}\right|l'\right\rangle$$

$$= (-)^{l+a+1}\left(\frac{a-1}{2a-3}\right)^{\frac{1}{2}}\left\langle l\left|\frac{d}{dr}+\frac{a}{r}\right|l'\right\rangle,$$

where the last factor is a radial integral and $a = -l'$ if $l = l'+1$, $a = l'+1$ if $l = l'-1$.

The matrices of tensor products and for composite systems are discussed in Chapter 5. Application of (5.9) gives

$$\langle l\tfrac{1}{2}j\|\mathbf{C}_k\|l'\tfrac{1}{2}j'\rangle = (2j'+1)^{\frac{1}{2}}(-)^{j'-k-\frac{1}{2}}\begin{pmatrix} j & j' & k \\ \frac{1}{2} & -\frac{1}{2} & 0 \end{pmatrix}$$

provided $l+l'+k$ is even, and zero otherwise;

$$\langle l\tfrac{1}{2}j\|\mathbf{T}_k(\mathbf{C}_k, \boldsymbol{\sigma})\|l'\tfrac{1}{2}j'\rangle = a_k(-)^{j'-k-\frac{1}{2}}(2j'+1)^{\frac{1}{2}}\begin{pmatrix} j & j' & k \\ \frac{1}{2} & -\frac{1}{2} & 0 \end{pmatrix}$$

where
$$a_k = (x-x')/\sqrt{\{k(k+1)\}},$$
$$a_{k-1} = -(k+x+x')/\sqrt{\{k(2k+1)\}},$$
$$a_{k+1} = (k+1-x-x')/\sqrt{\{(k+1)(2k+1)\}},$$

with
$$x = (l-j)(2j+1),\ x' = (l'-j')(2j'+1).$$

$$\langle j_1 j_2 J\|\mathbf{J}_1\|j_1' j_2' J'\rangle = \delta_{j_1 j_1'}\delta_{j_2 j_2'}(-)^{j_2+1-J-j_1}\times$$
$$\times[j_1(j_1+1)(2j_1+1)(2J'+1)]^{\frac{1}{2}}W(j_1 j_1 J J';\, 1 j_2),$$

and for example, since $\boldsymbol{\sigma} = 2\mathbf{s}$ for spin $\frac{1}{2}$

$$\langle l\tfrac{1}{2}j\|\boldsymbol{\sigma}\|l'\tfrac{1}{2}j'\rangle = \delta_{ll'}[\tfrac{3}{4}+j(j+1)-l(l+1)][j(j+1)]^{-\frac{1}{2}} \quad \text{if } j = j',$$
$$= \delta_{ll'}2[(l+1)/(2l+1)]^{\frac{1}{2}} \quad \text{if } j = j'-1$$

and
$$\langle \tfrac{1}{2}\tfrac{1}{2}S\|\boldsymbol{\sigma}_1\|\tfrac{1}{2}\tfrac{1}{2}S'\rangle = [S(S+1)]^{\frac{1}{2}} \quad \text{if } S = S',$$
$$= -\sqrt{3} \quad \text{if } S = S'-1.$$

Equations (5.12) and (5.13) give

$$\langle j_1 j_2 J \| \mathbf{J}_1 \cdot \mathbf{J}_2 \| j_1' j_2' J \rangle = \delta_{j_1 j_1'} \delta_{j_2 j_2'} \tfrac{1}{2}[J(J+1) - j_1(j_1+1) - j_2(j_2+1)],$$

$$\langle \tfrac{1}{2}\tfrac{1}{2}S \| \boldsymbol{\sigma}_1 \cdot \boldsymbol{\sigma}_2 \| \tfrac{1}{2}\tfrac{1}{2}S \rangle \quad = -3 \quad \text{if} \quad S = 0, \text{ and } 1 \quad \text{if} \quad S = 1.$$

Also $\langle \tfrac{1}{2}\tfrac{1}{2}S \| \mathbf{T}_2(\boldsymbol{\sigma}_1, \boldsymbol{\sigma}_2) \| \tfrac{1}{2}\tfrac{1}{2}S' \rangle = \delta_{SS'} \delta_{S1} (\tfrac{20}{3})^{\frac{1}{2}}$.

APPENDIX VII

ASYMPTOTIC EXPRESSIONS FOR LARGE ANGULAR MOMENTA AND CLASSICAL LIMITS

The classical 'vector model' has already been introduced in connection with the interpretation of the matrices for rotations of the coordinate system (p. 27), and of the vector addition, or Clebsch–Gordan, coefficients for the coupling of two angular momenta (p. 32). We would expect from the Correspondence Principle [46] that the various quantum mechanical expressions that appear in the treatment of angular momenta would reduce to the geometrical forms obtained from classical mechanics, when the magnitudes of the angular momenta become very large compared to $\hbar$. It should be remembered, however, that the reduction of quantum mechanics to classical mechanics may not converge uniformly. The existence of quantal interferences, which have no classical analogue, can result in the probability of observing a given value for some variable that oscillates as that value changes. These oscillations grow ever more rapid as the magnitude of the variable increases. Then, in order to see convergence to the smooth behaviour predicted classically, one must average the quantal oscillations over a small interval. We illustrate this property here for the spherical harmonics $Y_{lm}(\theta, \phi)$ when l is large; see also the discussion of the rotation matrices $\mathscr{D}^J_{M'M}$ on p. 29.

Another consequence of quantum uncertainties is that a variable may assume values that would fall outside the range allowable in a classical description. However, the probability of finding such a value decreases exponentially as the value moves further outside the classical range, and the decrease becomes more rapid as the magnitude of the variable becomes larger and more 'classical'. This phenomenon was mentioned on p. 29 in connection with the probability of finding a projection M' of a vector $\mathbf{J}$ on a new z'-axis after a rotation of the coordinate axes; other examples featuring spherical harmonics are illustrated by Zare [98]. Such nonclassical behaviour is closely analogous to another well-known quantal effect, the penetration of potential barriers [46, 99] that would be insurmountable in classical mechanics.

In this Appendix we give a few examples of asymptotic ('large j') expressions for some angular momentum coefficients and some alternate forms they may assume in this limit. Many others are listed in the exhaustive compendium of Varsholovich *et al.* [96], who give other referenes and sources. (See also the early discussions by Brussard and Tolhoek [15] and Wigner [78].) Some of these asymptotic expressions are surprisingly good for quite small values of the arguments. Besides illuminating the transition from the quantal to classical domains, asymptotic, large-j relations may find practical use in computations. Often one has the problem of evaluating sums over angular momenta that may take on very large values; this often occurs, for example, in descriptions of atomic, molecular, and heavy-ion collisions [97, 98]. The sums may be convertible into integrals after the use of one or more asymptotic relations. Sometimes the integrals may be performed analytically, thus greatly simplifying the results.

Spherical harmonics: Consider a particle which, classically, would be moving in a circular orbit with angular momentum of magnitude $[l(l+1)]^{1/2}\hbar \approx (l+\frac{1}{2})\hbar$ and z-component $m\hbar$. The orbit (Fig. A1) would be inclined to the z-axis at an angle $\beta = \sin^{-1}(m/[l(l+1)]^{1/2})$. The polar angle θ of the particle is limited to the range $\beta \leqslant \theta \leqslant \pi - \beta$.

In quantum mechanics, the x- and y-components of angular momentum become indeterminate, as though the orbit were precessing about the z-axis (p. 28), and the orbit is replaced by a probability distribution. This distribution is proportional to the

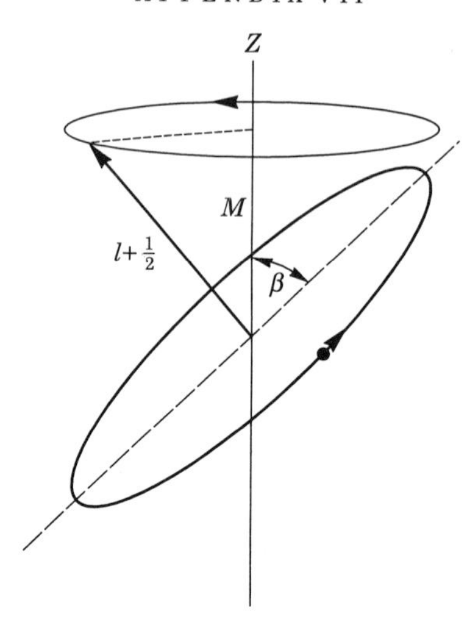

FIG. A1. Particle moving in a classical orbit with angular momentum $\sqrt{[l(l+1)]}\hbar \approx (l+\frac{1}{2})\hbar$ and z-component $m\hbar$. The plane of the orbit is inclined at angle β to the z-axis, where $\cos\beta = m/\sqrt{[l(l+1)]} \approx m/(l+\frac{1}{2})$. In the vector model, the vector l is supposed to precess around the z-axis.

square modulus of a spherical harmonic $|Y_{lm}(\theta, \phi)|^2$, which represents the probability of finding the particle with polar angles (θ, ϕ). The indeterminacy of the x- and y-components of angular momentum is represented by the uniform probability $d\phi/2\pi$ of finding the azimuthal angle between ϕ and $\phi + d\phi$.

The $m/l=0$ case is the easiest to visualize. Then the z-axis passes through a diameter of the corresponding classical orbit ($\beta=0$ in Fig. A1), while the plane of the orbit is distributed uniformly around the z-axis, with the probability $(2\pi)^{-1}$ rad^{-1}. Classically, it is easy to see that the probability of finding the particle with a polar angle in the interval θ to $\theta + d\theta$ is $d\theta/(2\pi^2 \sin\theta)$. Consequently, we would expect to find the same result in quantum mechanics in the limit of large l.

Indeed, there is an asymptotic form for the spherical harmonic, derived for $l \to \infty$ (but $l \gg m \geqslant 0$) which allows us to see this correspondence, namely

$$Y_{lm}(\theta, \phi) \approx \frac{e^{im\phi}}{\pi\sqrt{\sin\theta}} \sin\left[(l+\tfrac{1}{2})\theta + (2m+1)\frac{\pi}{4}\right], \qquad \text{(A.1a)}$$

provided θ is not too close to 0 or π ($\theta \gg l^{-1}$ or $\pi - \theta \gg l^{-1}$). We illustrate this for $m=0$, when (A.1a) becomes

$$Y_{l0}(\theta, \phi) \approx \frac{\sin\left[(l+\tfrac{1}{2})\theta + \dfrac{\pi}{4}\right]}{\pi\sqrt{\sin\theta}}. \tag{A.1b}$$

In fact, the 'large l' approximation (A.1b) is remarkably good for small l (even $l=1$!), provided $\theta > l^{-1}$. This is shown in Fig. A2 for $l=4$ and $l=10$, where the approximation (A.1b)is compared with the exact values of $Y_{l0}(\theta, \phi)$.

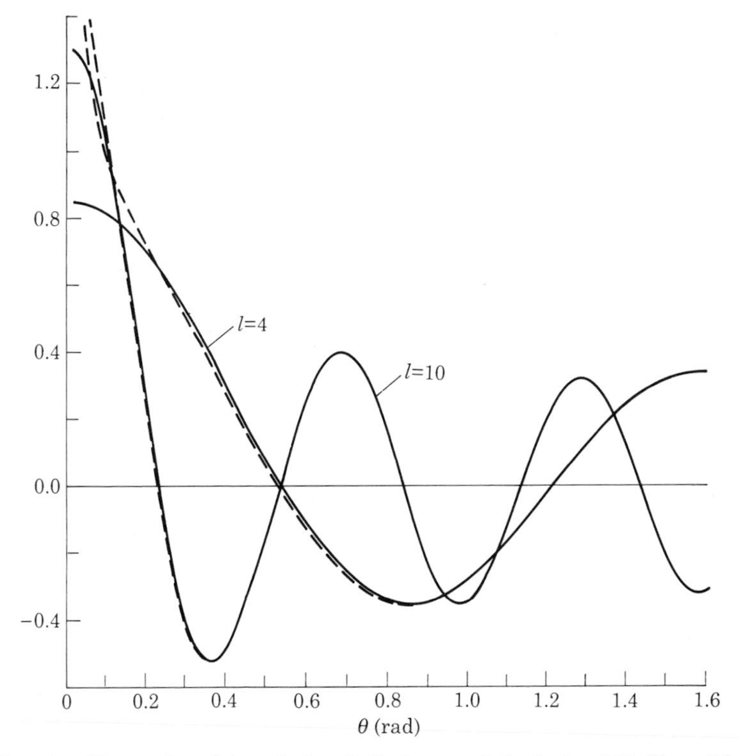

FIG. A2. Illustration of the relation (A.1) when $m=0$, for $l=4$ and 10. The solid curves represent the values of $Y_{l,0}(\theta, \phi)$, while the dashed curves are given by the approximation (A.1). Note that $Y_{l,0}(\pi-\theta, \phi) = (-)^l Y_{l,0}(\theta, \phi)$, and both are independent of ϕ.

When $\theta = \pi/2$, approximation (A.1b) predicts that

$$Y_{l0}\left(\frac{\pi}{2}, \phi\right) \approx \frac{(-)^l}{\pi}, \quad l \text{ even}$$

$$= 0, \quad l \text{ odd.} \tag{A.1c}$$

This is exact for odd l, and very close for even l. Even when $l = 2$, this estimate differs from the exact value by only one per cent.

When $m \neq 0$, the range of small θ for which (A.1a) is not a good approximation increases relative to that for $m = 0$. For example, the limit (A.1a) is not at all accurate for $Y_{10,2}(\theta, 0)$ until $\theta > \frac{1}{2}$, and for $Y_{10,4}(\theta, 0)$ until $\theta > 1$.

Classical limit for spherical harmonics: The last sine factor in the expression (A.1a) oscillates very rapidly with θ as l becomes large. This is an example of the quantal oscillations mentioned above. Consequently, the probability $|Y_{lm}(\theta, \phi)|^2$ must be averaged over an interval $\Delta\theta$ in θ which is small enough that the variation of the (sin θ) denominator may be ignored, but which contains many oscillations in the numerator; thus $\Delta\theta \gg \pi/l$. The the square of the oscillating sine factor may be replaced by its average value of $\frac{1}{2}$. (Note that the quantal oscillations persist however large l becomes, so that the averaging is always required, in principle, in order to make the transition from the quantum mechanical result to the corresponding smooth classical one.)

Thus, the averaged probability distribution becomes

$$|\overline{Y_{lm}(\theta, \phi)^2}| \approx (2\pi^2 \sin\theta)^{-1}, \tag{A.2}$$

provided $l \gg 1$, $l \gg m$, and θ is not too close to 0 or π. Since the condition $l \gg m$ implies $m/l \approx 0$, this approaches the classical case discussed above for $m/l = 0$. The approach to the correspondence (A.2) is illustrated in Fig. A3 where two examples of $|Y_{lm}(\theta, \phi)|^2$, with $m = 0$ and with $l = 10$ and $l = 25$, respectively, are compared to the classical expression $(2\pi^2 \sin\theta)^{-1}$. (See [98] for examples where $m \neq 0$.) The value of l is not very large in the

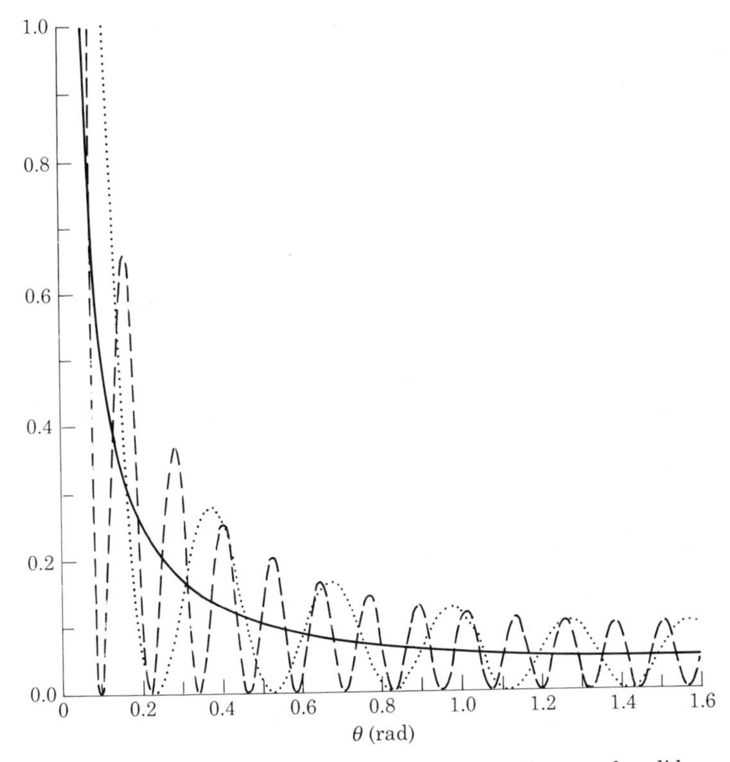

FIG. A3. Illustration of the relation (A.2) when $m=0$. The smooth, solid curve represents the approximation (A.2) while the oscillatory curves are values of $|Y_{l,0}(\theta, \phi)|^2$ for $l=10$ (short dashes) and $l=25$ (long dashes). Note that $|Y_{l,0}(\pi-\theta, \phi)|^2 = |Y_{l,0}(\theta, \phi)|^2$, and both are independent of ϕ.

former case, and an averaging interval $\Delta\theta \gg \pi/l = 0.3$ rad $= 18°$ would be required to smooth the oscillations. None the less, the average behaviour of $|Y_{10,0}(\theta, \phi)|^2$ follows the relation (A.2). The larger value $l=25$ would require a smaller averaging interval $\Delta\theta \gg 0.13$ rad $\approx 7°$, and the classical value (A.2) is approached more closely. None the less, the oscillations of $|Y_{lm}(\theta, \phi)|^2$ about the classical value continue, no matter how large l becomes; the spherical harmonic passes through zero at intervals of π/l in θ. Thus the averaging over θ is always needed *in principle* before we retrieve the classical limit (A.2), even though the averaging interval required becomes smaller as l becomes larger.

Another spherical harmonic limit: One can relate spherical harmonics to Bessel functions when $l \to \infty$ and $l \gg m \geqslant 0$, with $l\theta \gg 1$ but $\theta \ll 1$; then

$$Y_{l,-m}(\theta, \phi) \approx e^{-im\phi}\left(\frac{l+\frac{1}{2}}{2\pi}\right)^{1/2} J_m[(l+\tfrac{1}{2})\theta]. \quad \text{(A.3)}$$

When $m=0$, this reduces to

$$P_l(\cos\theta) \approx J_0[(l+\tfrac{1}{2})\theta]. \quad \text{(A.4)}$$

The asymptotic limits for the Bessel functions when $l\theta \gg 1$ show that the relations (A.3) and (A.4) are equivalent to (A.1) when θ is small.

These expressions can be used to show that equivalence of the Fraunhofer diffraction and partial-wave expansion descriptions of the scattering of a particle by a strongly absorbing ('black') disc or sphere [97, 99].

A generalization of the relation (A.3) holds for the reduced rotation matrix elements,

$$d^l_{mn}(\theta) \approx J_{m-n}[(l+\tfrac{1}{2})\theta], \quad \text{(A.5)}$$

again as $l \to \infty$, $l \gg m, n$, with $m-n \geqslant 0$, and $l\theta \gg 1$ but $\theta \ll 1$.

Vector addition (Clebsch–Gordan) coefficients: The interpretation of vector addition (Clebsch–Gordan) and other angular momentum coefficients as probability amplitudes, and their visualization in terms of the vector model, was discussed in Chapters II and III. It suggests that there are expressions for these quantities in the large-j limit that correspond to the geometrical relations that hold for the classical analogues.

Ponzano and Regge [100] (see also [96]) derived such expressions for the Clebsch–Gordan coefficient when all three angular momentum vectors are large, and showed that they have simple geometrical interpretations.

Another class of asymptotic expressions for Clebsch–Gordan coefficients arises when a small vector $\mathbf{b}$ is added to large one $\mathbf{a}$; consequently, the resultant $\mathbf{c}=\mathbf{a}+\mathbf{b}$ is also large. Two such expressions were given on p. 33, taken from the detailed discussion of Brussard and Tolhoek [15] (see also [96]). When $a,\ c \gg b$,

$$\langle ab\alpha\beta | c\gamma \rangle \approx \delta_{\gamma,\alpha+\beta}\, d^{b}_{\beta,c-a}(\theta) \tag{A.6}$$

where

$$\cos\theta = \gamma/\surd[c(c+1)] \approx \gamma/(c+\tfrac{1}{2}) \approx \gamma/c,$$

so that θ is the angle between the vector $\mathbf{c}$ and the z-axis. In the special case $a=c$ and $\beta=0$, this reduces to

$$\langle cb\gamma 0 | c\gamma \rangle \approx P_b(\cos\theta). \tag{A.7}$$

These approximations are quite accurate even when a and c are not very large. To illustrate this, consider the relation (A.7). Table 3 (p. 36) may be used to obtain explicit expressions for the Clebsch–Gordan coefficient when b is small. For example, if $b=1$, the left side of (A.7) becomes

$$\langle c1\gamma 0 | c\gamma \rangle = \gamma/\surd[c(c+1)],$$

which is *exactly* the right side for all values of c,

$$P_1(\gamma/\surd[c(c+1)]) = \gamma/\surd[c(c+1)].$$

(An error of order $1/8c^2$ would be introduced if the semiclassical approximation $\cos\theta = \gamma/(c+\frac{1}{2})$ were used.)

Again, if $b=2$, the Clebsch–Gordan coefficient is

$$\langle c2\gamma 0 | c\gamma \rangle = \frac{3\gamma^2 - c(c+1)}{\surd[c(c+1)(2c+3)(2c-1)]},$$

while the corresponding Legendre polynomial is

$$P_2(\gamma/\surd[c(c+1)]) = \frac{3\gamma^2 - c(c+1)}{2c(c+1)}.$$

The numerators are identical, while the denominators differ by relative amounts of order $(2c+1)^{-2}$.

n-j ($n>3$) coupling coefficients: Asymptotic expressions for coupling coefficients involving more than three angular momenta rapidly become more complicated and less easy to visualize as the number of variables increases, although there remain underlying geometrical structures that can be related to classical mechanics in the large-j limits. Various asymptotic relations for 6-j and 9-j symbols are included in the compilation of Varshalovich *et al.* [96]. Here we only present two examples for 6-j symbols (Racah coefficients).

The first is known as Edmonds' formula [22],

$$\begin{Bmatrix} abc \\ def \end{Bmatrix} \approx \frac{(-)^{a+c+d+f}}{\sqrt{[(2a+1)(2b+1)]}} d^f_{mn}(\theta), \qquad \text{(A.8)}$$

where $m = d - b$, $n = e - a$ and $a, b, c \gg 1, f, m, n$. The six vectors form a tetrahedron (Fig. A4), and θ is the angle between the vectors $\mathbf{a}$ and $\mathbf{b}$, so

$$\cos\theta = \frac{a(a+1) + b(b+1) - c(c+1)}{2\sqrt{[a(a+1)b(b+1)]}}. \qquad \text{(A.9)}$$

In the present case the vector $\mathbf{f}$ is short compared to the other vectors, and thus the differences m and n are small. In particular, if $m = n = 0$, relation (A.8) reduces to

$$\begin{Bmatrix} abc \\ baf \end{Bmatrix} \approx \frac{(-)^{a+b+c+f}}{\sqrt{[(2a+1)(2b+1)]}} P_f \cos\theta. \qquad \text{(A.10)}$$

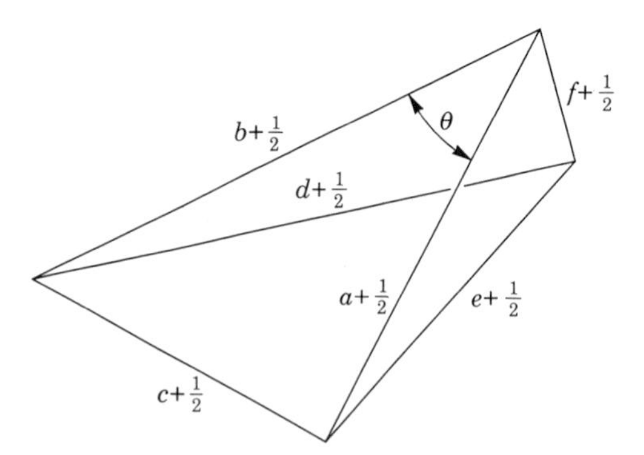

FIG. A4. Tetrahedron whose edges are formed by the angular momentum vectors represented by the arguments of the 6-j symbol $\begin{Bmatrix} abc \\ def \end{Bmatrix}$. The length of the vectors indicated makes use of the semiclassical approximation $\sqrt{[a(a+1)]} \approx a + \frac{1}{2}$, etc.

The other example is a simple expression, known as Wigner's formula [78], and applies when all six vectors which constitute the tetrahedron in Fig. A4 are large. If $a, b, c, d, e, f \gg 1$,

$$\begin{Bmatrix} abc \\ def \end{Bmatrix}^2 \approx \frac{1}{24\pi V}, \tag{A.11}$$

where V is the volume of the tetrahedron [78, 96]. However, the relation (A.11) is only valid on the average because the 6-j symbols oscillate rapidly in value as the arguments are varied. An average of the left-hand side of (A.11) over a reasonable range of values of at least one of the arguments must be taken before it converges to the right-hand side. This presents another example of the nonuniform convergence of quantal results to the classical expectations.

Extensions of the relations (A.8) and (A.11) are given by Varshalovich *et al.* [96].

REFERENCES AND AUTHOR INDEX

1. ALDER, K., BOHR, A., HUUS, T., MOTTELSON, B. and WINTHER, A. (1956) *Rev. Mod. Phys.* **28,** 432 98
2. ARIMA, A., HORIE, H. and TANABE, Y. (1954) *Prog. Theor. Phys.* **11,** 143 143
3. BANERJEE, M. K. and SAHA, A. K. (1954) *Proc. Roy. Soc. A* **224,** 472 151
4. BIEDENHARN, L. C. (1952) Oak Ridge National Laboratory, Report No. 1098 43
5. —— (1953) *J. Math. and Phys.* **31,** 287 . . 44, 56, 82, 132, 140
6. —— (1958) *Annals of Phys.* **4,** 104 . . . . . . 109, 110, 111
7. —— BLATT, J. M. and ROSE, M. E. (1952) *Rev. Mod. Phys.* **24,** 249 . . . . 19, 44, 140
8. —— and ROSE, M. E. (1953) *Rev. Mod. Phys.* **25,** 729 . . . . 94, 108, 109, 112
9. BLATT, J. M. and WEISSKOPF, V. F. *Theoretical Nuclear Physics* (John Wiley and Sons, Inc., New York, 1952). 11, 90, 136
10. BLEANEY, B. and STEVENS, K. W. H. (1953) *Reports on Progress in Physics* **16,** 108 . . . . 99, 100
11. BLIN-STOYLE, R. J. *Theories of Nuclear Moments* (Oxford University Press, 1957) 90, 95, 99
12. —— and GRACE, M. A. (1957) *Handbuch der Physik* **42,** 555 . . . . 93, 101
13. BOHR, A. and MOTTELSON, B. R. (1953) *Dan. Mat. Fys. Medd.* **27,** No. 16 . . . . 27, 90, 104, 146
14. BRINK, D. M. and SATCHLER, G. R. (1956) *Nuovo Cimento* **4,** 549 . . . . 83, 96
15. BRUSSARD, P. J. and TOLHOEK, H. A. (1957) *Physica* **23,** 955 . . . . 29, 33, 155, 161
16. BUEHLER, R. J. and HIRSCHFELDER, J. O. (1951) *Phys. Rev.* **83,** 628 56
17. CONDON, E. U. and SHORTLEY, G. H. *Theory of Atomic Spectra* (Cambridge University Press, 1935) viii, 17, 18, 83, 96, 104, 136, 145
18. DE-SHALIT, A. (1953) *Phys. Rev.* **91,** 1479 35
19. DEVONS, S. and GOLDFARB, L. J. B. (1957) *Handbuch der Physik* **42,** 362 . . . . 94, 108, 109
20. DIRAC, P. A. M. *Principles of Quantum Mechanics* (Oxford University Press, 1947) 31, 77, 111, 149
21. ECKART, C. (1930) *Rev. Mod. Phys.* **2,** 305 58
22. EDMONDS, A. R. *Angular Momentum in Quantum Mechanics* (Princeton University Press, 1957) . . . . viii, 6, 21, 57, 60, 113, 136, 146, 149, 151, 162
23. ELLIOT, J. P. (1953) *Proc. Roy. Soc. A* **218,** 345 . . . . 44, 106, 107, 132, 140
24. —— and FLOWERS, B. H. (1955) *Proc. Roy. Soc. A* **229,** 536 . 47
25. —— JUDD, B. R. and RUNCIMAN, W. A. (1957) *Proc. Roy. Soc. A* **240,** 509 99
26. —— and LANE, A. M. (1957) *Handbuch der Physik* **39,** 241 . . . . 83, 86, 101, 102
27. FALKOFF, D. L. and UHLENBECK, G. E. (1950) *Phys. Rev.* **79,** 323 149

28. FANO, U. (1951) National Bureau of Standards, Report No. 1214 . . . . 45, 109
29. —— (1953) *Phys. Rev.* **90,** 577 109
30. —— (1957) *Rev. Mod. Phys.* **29,** 76 108
31. —— and RACAH, G. *Irreducible Tensorial Sets* (Academic Press, Inc., New York, 1959) . v, 21, 57, 108, 109, 136, 146, 151
32. FORD, K. W. and KONOPINSKI, E. J. (1958) *Nuclear Physics* **9,** 218 104
33. GELL-MANN, M. and GOLDBERGER, M. L. (1953) *Phys. Rev.* **91,** 398 112
34. HEITLER, W. *Quantum Theory of Radiation* (Oxford University Press, 1954) 71, 88
35. HILL, E. L. (1954) *Am. J. Phys.* **22,** 211 146
36. HOPE, J. and LONGDON, L. W. (1956) *Phys. Rev.* **101,** 710 . 107
37. —— and —— (1956) *Phys. Rev.* **102,** 1124 106
38. HOWELL, K. M. (1959) University of Southampton, Research Report No. 59–1 . . . . 45, 140
39. JAHN, H. A. and HOPE, J. (1954) *Phys. Rev.* **93,** 318 47, 136, 143
40. JAHNKE, E. and EMDE, F. *Tables of Functions* (Dover Publications, New York, 1945) 18, 145
41. KUSCH, P. and HUGHES, V. W. (1959) *Handbuch der Physik* **37,** 1 . . . . 92, 98, 99
42. LANDAU, L. D. and LIFSCHITZ, E. M. *Quantum Mechanics* (Pergamon Press, London, 1958) 136, 145
43. LANE, A. M. and RADICATI, L. A. (1954) *Proc. Phys. Soc. A* **67,** 167 96
44. —— and THOMAS, R. G. (1958) *Rev. Mod. Phys.* **30,** 257 . 112
45. LEE-WHITING, G. E. (1958) *Can. J. Phys.* **36,** 1199 . . . 151
46. MESSIAH, A. *Mechanique Quantique* (Dunod, Paris 1, 1960 Ch XIII . . . . viii, 6, 21, 146, 154
47. ORD-SMITH, R. J. (1954) *Phys. Rev.* **94,** 1227 47
48. RACAH, G. (1942) *Phys. Rev.* **62,** 438 . . 34, 41, 43, 57, 140, 151
49. —— (1943) *Phys. Rev.* **63,** 367; ibid **76,** 1352 . . . . 83, 96
50. RAMSEY, N. F. *Molecular Beams* (Oxford University Press, 1955). 90, 99
51. REGGE, T. (1958) *Nuovo Cimento* **10,** 544 136
52. —— (1959) *Nuovo Cimento* **11,** 116 140
53. ROSE, M. E. *Multipole Fields* (John Wiley and Sons, Inc., New York, 1955) 21, 88, 98, 146
54. —— *Elementary Theory of Angular Momentum* (John Wiley and Sons, Inc., New York, 1957). . viii, 6, 21, 57, 88, 90, 136, 146
55. —— *Internal Conversion Coefficients* (North Holland Publishing Co., Amsterdam, 1958) 98
56. —— (1958) *J. Math. and Phys.* **37,** 215 63
57. —— and OSBORN, R. K. (1954) *Phys. Rev.* **93,** 1326 . . . 151
58. ROTENBERG, M., BIVINS, R., METROPOLIS, N. and WOOTEN JR. J. K. *The 3-j and 6-j Symbols* (Technology Press, M.I.T., Camb., Mass., 1959) 35, 43, 45
59. SCHWARTZ, C. (1955) *Phys. Rev.* **97,** 380 . . 77, 88, 92, 98, 99
60. SCHWINGER, J. (1952) U.S. Atomic Energy Commission, Report No. NYO-3071 60

61. SHARP, W. T. (1955) Atomic Energy of Canada Ltd., Report No. TPI-81 . . . . 47, 143
62. —— (1956) *Bull. Am. Phys. Soc.* **1,** 210
63. —— (1957) Atomic Energy of Canada Ltd., Report No. CRL-43 90
64. SHORTLEY, G. H. and FRIED, B. (1938) *Phys. Rev.* **54,** 739 . . 104
65. SIMON, A. (1954) Oak Ridge National Laboratory, Report No. 1718 136
66. SIMON, A., VANDER SLUIS, V. H. and BIEDENHARN, L. C. (1954) Oak Ridge National Laboratory, Report No. 1679 35, 43
67. SMITH, K. and STEVENSON, J. W. (1957) Argonne National Laboratory, Reports No. 5776, 5860-I, and 5860-II . . . 47
68. STEVENS, K. W. H. (1952) *Proc. Phys. Soc. A* **65,** 209 . 99, 100
69. SWIATECKI, W. (1951) *Proc. Roy. Soc. A* **205,** 238 102
70. TALMI, I. (1952) *Helv. Phys. Acta* **25,** 185 . . . . 102, 104
71. —— (1953) *Phys. Rev.* **89,** 1065 107
72. TER HAAR, D. *Elements of Statistical Mechanics* (Rinehart and Co., New York, 1954) 108
73. THIEBERGER, R. (1957) *Nuclear Physics* **2,** 533 104
74. WANG, T. C. (1955) *Phys. Rev.* **99,** 566 92
75. WEYL, H. *Classical Groups* (Princeton University Press, 1939 . . . . 23, 149
76. WIGNER, E. P. (1937) *Phys. Rev.* **51,** 106 106
77. —— (1937) *On the Matrices which Reduce the Kronecker Products of Representations of S.R. Groups* (unpublished) . . . . viii, 44, 45, 47
78. —— *Group Theory and its Application to the Quantum Mechanics of Atomic Spectra* (Academic Press, New York, 1959). . . . . viii, 6, 11, 21, 23, 38, 57, 58, 146, 155, 163
79. WILETS, L. (1954) *Dan. Mat. Fys. Medd.* **29,** No. 3 . . . 98
80. WINTER, C. VAN (1954) *Physica,* **20,** 274 27

SUPPLEMENTARY REFERENCES

81. BIEDENHARN, C. L. and VAN DAM, H. *Quantum Theory of Angular Momentum* (Academic Press, New York, 1965) . . viii
82. DÖNAU, F. and FLACH, G. (1965) *Nucl. Phys.* **69,** 68 . . . 113
83. EL-BAZ, E., MASSOT, J. N. and LAFOUCRIÈRE, J. (1966) *Nucl. Phys.* **82,** 189; MASSOT, J. N., EL-BAZ, E. and LAFOUCRIÈRE, J. (1966) *Nucl. Phys.* **83,** 449; EL-BAZ, E., MASSOT, J. N. and LAFOUCRIÈRE, J. (1966) *Nucl. Phys.* **86,** 625; (1967) *Rev. Mod. Phys.* **39,** 288 113
84. JUDD, B. R. *Operator Techniques in Atomic Spectroscopy* (McGraw-Hill, New York, 1963) 113
85. LEVINSON, J. B. (1957) *Trudy fiz.-tekh. Inst., Ashkhabad,* **2,** 17, 31; (1957) *Liet. TSR Mokslu Akad. Darb.* B**4,** 3 . . . 112
86. YUTSIS, A. P., LEVINSON, J. B. and VANAGAS, V. V. *Mathematical Apparatus of the Theory of Angular Momentum.* Israel Program for Scientific Translation (Jerusalem, 1962) . . 113
87. —— and BANDZAITIS, A. A. *Quantum Theory of Angular Momentum* (Moscow, 1965) 113

88. Wolf, A. A. (1969) *Am. J. Phys.* **37,** 531 vi
89. Bouten, M. (1969) *Physica* **42,** 572 vi
90. Wolters, G. F. (1970) *Nuclear Physics* **B18,** 625 vi
91. Buckmaster, H. A. (1964) *Can. J. Phys.* **42,** 386; (1966) *Can. J. Phys.* **44,** 2525 vi
92. Way, K. and Hurley, F. W. (1966) *Nuclear Data Tables* **A1,** 473; see also cumulated subject index in *Nuclear Data Tables* **A8,** 670 (1971) vi
93. Sharp, R. T. (1967) *Nuclear Physics* **A95,** 222 vi
94. Rose, H. J. and Brink, D. M. (1967) *Rev. Mod. Phys.* **39,** 306 . vi
95. Rybicki, F., Tamura, T., and Satchler, G. R. (1970) *Nuclear Physics* **A146,** 659 vi
96. Varshalovich, D. A., Moskalev, A. and Khersonskii, V. K. *Quantum Theory of Angular Momentum* (World Scientific, Singapore, 1988) 33, 160, 162, 163
97. Brink, D. M. *Semiclassical Methods in Nucleus–Nucleus Scattering* (Cambridge University Press, 1985) 155, 160
98. Zare, R. N. *Angular Momentum* (Wiley-Interscience, New York, 1988) . . . . v, 155, 158
99. Satchler, G. R. *Introduction to Nuclear Reactions* (Macmillan, London, 1990) 160
100. Ponzano, G. and Regge, T. in *Spectroscopic and Group Theoretical Methods in Physics* (North-Holland, Amsterdam, 1968) 160

SUBJECT INDEX